NUFFIELD DESIGN & TECHNOLOGY
STUDENT'S BOOK

LONGMAN

Longman Group Ltd
Longman House, Burnt Mill, Harlow,
Essex CM20 2JE
and associated companies throughout the world.

First published 1995
Fourth impression 1995
ISBN 0582 21266 9

© Nuffield Foundation 1995

All rights reserved; no part of this publication may be reproduced, stored in a retrieval system, or transmitted in any form or by any means, electrical, mechanical, photocopying, recording or otherwise without either prior written permission of the Publishers or a licence permitting restricted copying issued by the Copyright Licencing Agency Limited, 90 Tottenham Court Road, London, W1P 9HE.

Set in Garamond ITC Light Bt 12/13
Designed and produced by Gecko Ltd, Bicester, Oxon OX6 0JT.
Printed in Great Britain
by Butler & Tanner Limited, Frome and London

The publisher's policy is to use paper manufactured from sustainable forests.

Project directors

Executive director Dr David Barlex (Goldsmiths' College, University of London)

Co-directors Prof. Paul Black (King's College, London) and Prof. Geoffrey Harrison (Nottingham Trent University)

Deputy director – dissemination David Wise (Windlesham House School)

Contributors

Dr David Barlex, Goldsmiths' College, University of London
Catherine Budgett-Meakin, Intermediate Technology Group
Jo Compton, Institute of Education, University of London
Keith Everett, Education Consultant
David Fair, Thames Television
Nick Given, Exeter University
Ann Hampton, Intermediate Technology Group
Margaret Jepson, John Moores University, Liverpool
Niel McLean, Ealing TVEI
Mike Martin, Intermediate Technology Group
Jane Murray, Somerset LEA
John Plater, Old Palace School, Croydon
Val Rea, Intermediate Technology Group
Marion Rutland, Roehampton Institute
Eileen Taylor, Sefton LEA

Early developments

Ian Fletcher, Warwickshire LEA
Ann Hepher, Kingston LEA
School of Education, University of Bath
South West Region School Technology Forum
North West Region of TVEI
Nottingham Technology Education Development Group, Nottingham Trent University
Nuffield Chelsea Curriculum Trust
United Biscuits Leicestershire Development Group

The Nuffield Design and Technology Project gratefully acknowledges the support of the following commercial concerns in developing the published materials:

GEC Education Liaison

Zeneca Pharmaceuticals (formerly ICI Pharmaceuticals)

Tesco Training & Education Department

United Biscuits UK Ltd

Project manager Diona Gregory

Editors Katie Chester, Helen Johnson, Bettina Wilkes

Picture researchers Louise Edgeworth, Penni Bickle, Charlotte Deane, Marilyn Rawlings

Indexer Richard Raper/Indexing Specialists

Illustrations by Gecko Ltd, Barry Atkins, Gill Bishop, Joe Little, John Plumb, Archie Plumb, Martin Sanders, Tony Wilkins

Safety

Longman Education is grateful to Martin Trevor for advice on safety matters. Every effort has been taken to ensure safe practice.

Contents

Using this book	1

1 Meeting needs and wants — 2

Needs wants and differences — 2
Physical variation — 4
The natural environment — 6
Personal appearance — 8
Warmth — 12
Shelter — 14
Food and drink — 16
Carrying — 20
Safety — 22
Recreation — 25
Interior design — 28

2 Strategies — 30

Working methods for designing — 30
Identifying needs and likes — 31
Design briefs — 36
Specifying the product — 38
Generating design ideas — 40
Modelling – how it can help — 44
Modelling product appearance on paper — 45
Modelling product appearance in 3D — 50
Modelling product performance on paper — 52
Modelling product performance in 3D — 56
Modelling with computers — 62
Using a systems approach — 66
Planning techniques — 70
Evaluating outcomes — 72

3 Communicating design ideas — 76

Why and how? — 76
Presenting your product idea — 78
Presentation drawings for textile designs — 82
Presenting food product ideas — 84
Presenting your idea in a context — 85
Displays and exhibitions — 87
Presentation reports and documents — 88
Using facts and figures in presentations — 90
Information for making — 92
User support — 96

4 Designing and making with mechanisms — 99

What can mechanisms do? — 99
Using wheels and axles — 102
Using shafts, bearings and couplings — 103
Using gears — 104
Using pulleys and sprockets — 107
Using cranks, levers and linkages — 110
Using cams, eccentrics, pegs and slots — 114
Using screw threads — 117
Mechanisms Chooser Chart — 118
Using springs — 120
Using syringes — 121

5 Designing and making with textiles — 123

Where do textiles come from? — 123
What are textiles like? — 124
Fabrics Chooser Chart — 124
Testing fabrics — 125
Explaining choices — 126
Preparing and caring for fabrics — 128
Construction techniques — 129
Construction techniques – seams — 130
Construction techniques – edges and edging — 132
Construction techniques – cords and elastic — 134
Construction techniques – fastenings — 135
Fastenings Chooser Chart — 135
Ways to decorate fabrics — 136

Fabric Decoration Chooser Chart	137
Introducing patterns	138
Designing glove puppets	140
Designing soft toys	141
Designing bags	142
Designing garments from rectangles	144
Designing shorts	146
Designing T-shirt tops	148
Designing hats	150

6 Designing and making electric circuits — 152

Batteries	152
Making connections	153
Components to make things happen	154
Using switches	156
Designing electric circuits	158
Electric Components Chooser Chart	160

7 Designing and making electronic circuits — 162

Electronic control	162
Sensors and processors	163
The relay	167
Making a printed circuit board (PCB)	168
Designing an electronic product	170
Sensing with Electronics Chooser Chart	172

8 Computer control — 174

Computer control systems	175
Programs	175

9 Understanding structures — 178

The arch bears up	178
Aloft on the aerial ropeway	179
Walking the plank	180
Upsetting the apple-cart	181
Step-ladders that won't stay up	182
Strength in hollow boxes	183

10 Designing and making with resistant materials — 184

Choosing resistant materials	184
Resistant materials information	186
Solid Timber Chooser Chart	186
Manufactured Board Chooser Chart	187
Metals Chooser Chart	188
Plastics Chooser Chart	189
Work on it safely	190
Marking out and checking that it's right	192
Cutting the pieces and trimming them	194
Making holes	197
Forming with resistant materials	199
Casting	202
Using machine tools	203
Joining	204
Adhesives Chooser Chart	204
Fittings Chooser Chart	205
Assembling different parts	208
Making resistant materials look good and stay good	209
Finishes Chooser Chart	213
Choosing tools	214
Tools Chooser Chart	214

11 Designing and making with food — 216

The food materials available	216
The properties and qualities of food	218
Food choices	220
Designing food products	224
Food Wrappings Chooser Chart	229
Preparing food ingredients	231
Combining different food materials	233
Cooking the food	234
Finishing touches	238
Sensory evaluation tests	239
Prolonging shelf-life	242
Careful practice and food safety	244

Glossary and Index — 245

Using this book

When you do design and technology at school you have to design and make things for people to use. This can be tricky and there will be times when you...

- are puzzled;
- can't work out what's important;
- don't know how to do something;
- need some information;
- can't decide what will work.

That's when you need this *Student's Book*. It contains lots of information that is useful for design and technology.

To help you understand and use this information it is linked to short practical activities called Resource Tasks. The reference numbers for the tasks are given in boxes like this:

Resource task
SRT 1

1 Meeting needs and wants

Needs, wants and differences

What do we really need?

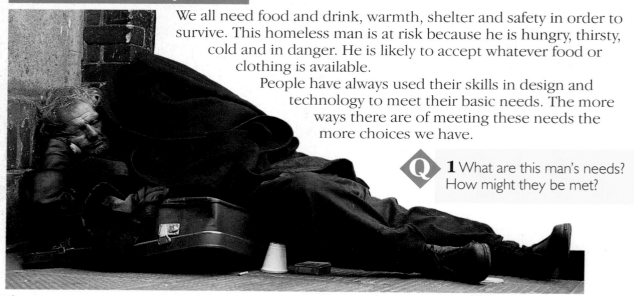

We all need food and drink, warmth, shelter and safety in order to survive. This homeless man is at risk because he is hungry, thirsty, cold and in danger. He is likely to accept whatever food or clothing is available.

People have always used their skills in design and technology to meet their basic needs. The more ways there are of meeting these needs the more choices we have.

Q 1 What are this man's needs? How might they be met?

▲ *This man has little choice in how to meet his needs.*

What do we want?

The shopper has chosen how she will meet her needs for food and drink. Her choice is formed by:
- the number of people she has to feed;
- the money available;
- her personal preferences;
- her cultural background;
- advertising.

Q 2 How might the shopper's cultural background affect what she wants? How might advertising affect what she wants?

▲ *This shopper has a wide range of choices.*

The same but different

Although we share the same basic needs, each one of us is different from everyone else. This difference shows itself in many ways.

There is our physical appearance (see the next section). Then there is our personality. A person may be a loner or sociable, talkative or quiet. Our tastes also differ. Some of us prefer sweet- to savoury-tasting foods – or bright primary colours to more delicate shades.

We often show our individuality by making our personal possessions look different. You probably decorate your pencil case and the covers of your books or your bedroom to reflect your tastes and interests.

▲ *Each person at this market stall will probably buy something different.*

This is what I like

The way we choose to spend our leisure time and the things we choose to buy also reflect our individuality. What we do will depend on how much money we have and what we like.

Designers who create goods for shops try to appeal both to our individual tastes and to a mass market.

 3 Think of something that you have changed to make it personal to you:
a Why did you do it?
b How did you do it?
c What difference did it make?
d Were you pleased with the result?

▲ *Tastes in pop music vary widely.*

 4 Find out about the range of tastes in your class for:
a snack foods;
b drinks;
c music;
d pens and pencils.

Find out what each person likes best and what they like least. Design a short questionnaire and put the results in a table.

1 MEETING NEEDS AND WANTS

Physical variation

Different shapes and sizes

People come in many shapes and sizes. Designers need to take this into account so that the things they design are suitable for individual users. Scientists help designers by collecting information (data) about people's shapes and sizes. This is called **anthropometric** data.

Designers also need to know about **ergonomics**. This gives useful information for designing equipment or a working environment that will be comfortable and efficient to use or work within.

 1 Measure the circumferences of the heads of the people in your class. Present the data you collect:
a as a table;
b as a graph;
c as a series of models.

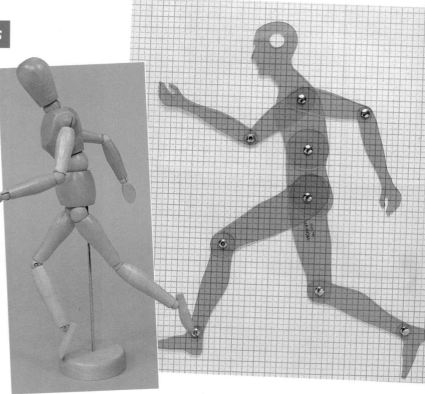

▲ *Sometimes designers use physical models like these 'ergonomes', rather than tables of data, to help them work out a design.*

Designed to be held

We use our hands whenever we use tools or equipment to grip tightly or adjust delicately. Hands come in many different sizes. The design of things that have to be held, touched or moved by our hands is a challenge to the designer.

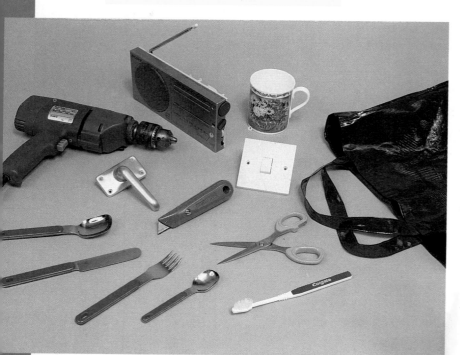

▲ *You might use all of these 'handles' in a single day. The way you use them will depend on how easy they are to hold, touch or grip.*

 2 Sort the 'handles' in the figure into those:
a gripped tightly;
b held or touched lightly.
Sort each of these into those that are:
c turned or twisted;
d pushed or pulled.
Can you spot any patterns in the design of the different sorts of handles?

Designed to fit

The ready-to-wear clothes that we buy in high street shops are mass-produced in a range to fit people of different sizes. Some clothes, such as T-shirts, only have to fit approximately so they are produced in a few sizes. Shoes must fit much more exactly so they are produced in a wider range of sizes and are often made adjustable by means of laces or straps for a better fit.

3 Find out the variation in shoe size in your class. Present the data you collect:
a as a table;
b as a graph;
c as a diagram showing the difference between the largest and smallest shoe size.

▲ *T-shirts and trainers have different size requirements.* ▶

Designed around us

A working **environment** needs to be designed to prevent workers from tiring easily or suffering stress.

In the kitchen shown here someone may work several hours daily. The designer needs to consider these points:

- If the shelves are too high, the person has to stretch to reach what is on them.
- If the worktop is too low, the person will have to stoop slightly and may get backache.
- If the door handle on the oven is too low, the person has to bend more than is necessary.
- If the control panel for the cooker is poorly laid out, it will take longer than necessary to set the cooking conditions.
- If the storage, sink and draining board, work surfaces and equipment are poorly organized, each operation will take longer than necessary.

Designers produce working environments that are suitable for *most* people, but those who have special needs may find working in these conditions difficult.

▲ *This kitchen may not be a good working environment.*

The natural environment

A planet at risk

If we damage our natural environment so much that the Earth becomes a dangerous place on which to live, we will all suffer. How we use design and technology seriously affects our natural environment.

Global warming

When large quantities of fossil fuels (oil, coal and gas) are burned they produce carbon dioxide, so the amount of this gas tends to increase in the atmosphere.

This gas traps heat from the Sun, making the world hotter and causing droughts which will make farming more difficult. Widespread flooding may result from the rise in sea level as polar ice caps melt.

Deforestation

Large areas of the Earth's natural forest are being cleared to provide timber or pulp for making paper, or to make room for farming. New trees are often single varieties that do not support the varied life of the old forest. Many species of plants and animals die out when old forests are cut down.

Trees use carbon dioxide in photosynthesis, so forests help control the level of carbon dioxide in the atmosphere. As the forests disappear this control is lost and global warming increases.

▲ *Electricity from this coal-fired power station is useful but the price includes global warming.*

▲ *The chain-saw has made forest clearing quick and easy.*

 1 What else is produced using fossil fuels? What proportion of each is burned or turned into materials used to make things?

▲ *Forests are beautiful as well as important.*

Choices

The way we choose to use design and technology in the future could solve environmental problems. Here are some possibilities.

Renewable energy sources

Wind turbines could be used to generate some of our electricity. This reduces the amount of fossil fuel we use. Other possibilities include tidal and wave energy.

▲ *A wind-energy farm.*

Designed to be thrown away?

Lots of the packaging on products we buy is made of paper produced from wood pulp. It makes the products look attractive and protects them. In the end we just throw it all away. Using recycled materials for packaging and making packaging that is easy to recycle are two ways to save trees.

Disposable nappies are made from paper pulp and make up 4 per cent of British household waste. One alternative is washable traditional cotton nappies. But the energy cost of washing the nappies and the damage the detergent might do to the environment need to be considered. Using recycled paper for disposable nappies might be a good alternative.

▲ *Disposable – an alternative to washable nappies.*

Q2 List paper-based products that could be made out of recycled paper. Can you think of any problems?

Designed to be recycled

A car contains many different materials. Designing it so that the materials can be recovered and recycled is a major challenge to designers. One manufacturer is building special dismantling plants for a popular make of car so that each car can be recycled at the end of its useful life.

Q3 Collect the packaging from the products you buy during a week. Sort it into metal, plastic, paper, glass and mixed, then make a display on recycling.

▲ *This car was designed to be recycled.*

Personal appearance

What makes up our personal style?

The factors that make up our personal style and appearance include clothes, footwear, hairstyle, make-up and body adornment. Examples are shown in the panel below. In developing your own personal style or designing one for someone else you will need to consider each of these important areas.

Key features of personal style

Hair
Hair can be cut or left to grow, curled, straightened, coloured and stiffened. It can be tied back or left free.

Make-up
We use make-up to protect our skin, hide blemishes, decorate ourselves and to draw attention to particular parts, such as eyes and lips.

Footwear
Footwear was originally developed to protect our feet and make walking easier, but shoes are now important items of personal style.

1 Describe your personal style in terms of the five areas in the panel. Are any other areas important for your personal style?
2 Start a scrapbook showing different personal styles. Add notes on each of the areas described here.

Body adornment
We use body adornment to decorate ourselves, to celebrate an event or occasion, to make ourselves look attractive, to draw attention to ourselves, or to show that we belong to a particular group. It can be painful and dangerous.

Clothing
Clothing is an important part of our personal style. Formal clothing or uniforms may show clearly who or what a person is or does. Casual clothing gives few clues.

...Personal appearance

Other times and cultures – illustrations to help you design

These two pages contain illustrations of styles from around the world and times past. Use them as a source of design ideas in, for example:

- designing clothing;
- designing jewellery;
- designing patterns (for textiles or wallpaper);
- suggesting shapes for logos and symbols;
- suggesting shapes and decoration for containers.

Answering question **1** will help you to understand the different styles and use them in your designing.

1 Ask yourself these questions about each style:
  a What colours are important?
  b What shapes are important?
  c What patterns are important?
  d What materials are important?
  e Do you find anything unusual or odd about the style? If so, what?
  f How does the style make you feel: happy or sad, threatened or safe, anything else?
  g Do you find one style more attractive than the rest? Explain why.

1 MEETING NEEDS AND WANTS

Warmth

Too hot, too cold!

Our bodies need to stay at a constant temperature of about 37°C. When we are ill, it sometimes goes a degree or two higher or lower. If it goes several degrees higher or lower we become very ill and could die. People have always designed and made things to help keep their bodies at the right temperature.

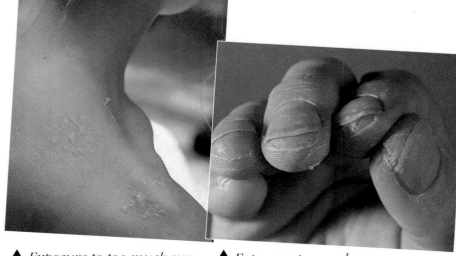

▲ *Exposure to too much sun causes sunburn.*

▲ *Exposure to very low temperatures causes frost-bite.*

 1 What have people designed and made to prevent sunburn?
2 What have people designed and made to prevent frost-bite?
3 What can be done to avoid these problems?

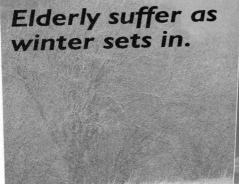

▲ *Extremes of temperature cause problems for many people.*

 4 What materials have been used for the clothes in the pictures, and why were they chosen? How do the designs meet the requirements listed?

Keeping warm outdoors

Our clothes must keep us warm, and also:
- leave us free to move;
- be comfortable;
- resist wear and tear;
- look attractive;
- be made from available materials.

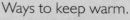

Ways to keep warm.

Keeping warm indoors

Indoor clothing is usually lightweight. But to keep warm indoors we usually heat our homes rather than wear extra clothes.

We can use small heaters or fires for individual rooms, or a central heating system for the whole house.

 5 List advantages and disadvantages of each heating method shown in the figure.

We **insulate** our homes to keep the heat in, stop draughts and save fuel.

6 What materials have been used to insulate your house?

▲ *Four ways to heat a room.*

What about keeping cool?

Overheating is just as serious as getting too cold.

It is refreshing to have either a cold or a hot drink when we are hot. Refreshing drinks are useful because they replace water lost through sweating.

Breezes keep us cool; artificial breezes include hand-operated fans, electrically driven fans and air-conditioning systems.

People who live in hot countries have designed clothes that help keep them cool.

▲ *Ways of keeping cool.*

 7 What do you drink to cool down? Is it hot or cold?

8 How do you cool a hot room at home or at school?

9 What clothes do you wear to keep cool?

Shelter

Shelter from the elements

People design and make structures for shelter. These structures must be strong and able to keep their shape to stand against the wind. The covering needs to be draught-proof and waterproof. The entrance should be designed so that it does not spoil this comfort. Sometimes the structure needs to provide shade from the Sun.

If the shelter is used by people who are always on the move then it has to be taken down and put up quickly and easily. If it is for people who are settled, it will be built on foundations that keep it firmly fixed to the ground.

▲ *A hurricane shows the power of the elements.*

In the desert

The Bedouin are nomadic people who live in the deserts of Arabia. They have developed a tent that provides shade during the day, keeps in warmth at night and can withstand fierce sandstorms. They use a framework covered by fabric which is held down by ropes that are pegged into the ground.

 1 Study the Bedouin tent in the picture.
  a Sketch the frame and ropes that form its supporting structure.
  b Label the parts that resist **compression** forces and those that resist **tension**.
  c How does the framework fit together?
  d Sketch the shape of the fabric that covers the framework.
  e Explain how the structure of the tent holds the fabric in the right shape.

▲ *The Bedouin tent provides shelter suitable for both day and night.*

Waiting for a bus

A bus shelter poses an interesting design problem because of the different needs it has to fill. It must:

- provide shelter from wind and rain – this requires a closed-in structure;
- allow people to see the bus coming and to leave the shelter quickly and easily to stop the bus – these require an open structure;
- be firmly fixed to the ground, solid, vandal-proof and easy to recognize, but it should not spoil its surroundings;
- provide weather-proofed timetable information.

The design which will fill all these conflicting needs will be a compromise.

▲ *Bus shelters have to meet conflicting needs.*

 2 Draw a sketch of the bus shelter in the figure and label the parts:

a that give shelter;
b that allow visibility;
c that allow easy access.
Does it look attractive?

3 Look at the bus shelters in your area to see how well they deal with the conflicting needs. Try to work out why they might have been set up in their sizes and positions.

 4 Sketch the stadium shown in the figure and label the parts concerned with:

a shelter;
b seating or standing room;
c crowd control.

Mark in on your sketch how spectators might enter and leave the stadium.
How would you improve the design?

Open-air sports

An open-air sports stadium is just a very large shelter which also has to fill a range of sometimes conflicting needs:

- It has to keep wind and rain off the spectators while allowing them to watch outdoor sports.
- The design must allow large numbers of people to enter and leave safely.
- It must provide a place for spectators to sit or stand and prevent them from running onto the pitch.
- When spectators are stamping and jumping up and down with excitement, the design must be strong enough for the structure not to break or work loose from the fixings and collapse.

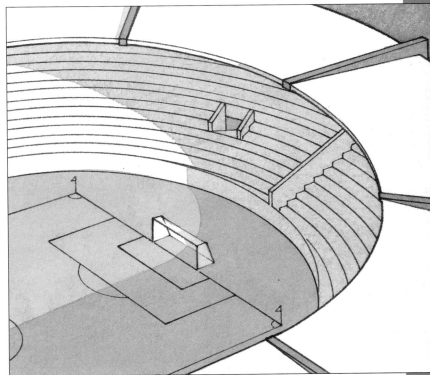

▲ *The design of an open-air sports stadium.*

Food and drink

Producing the food we need

People everywhere have always used design and technology to get enough to eat.

Food is produced on farms all over the world. Some farms are very small and the work is done mainly by hand with simple tools. The farm may produce little more than the food needed to support those working the land. Other farms are very large and highly mechanized. They produce very large quantities of food which may be sold across the world.

The basic process is the same for both small- and large-scale farmers. Both must sow their crops, tend them and then harvest them. They must sell their crops for enough to support themselves and buy more seed for the next year.

1 Compare small and large farms under the following headings:
  a the equipment used, and its maintenance and running costs;
  b storage needed;
  c how the produce is sold;
  d the effects of drought and crop disease.

▲ *Small- and large-scale farming.*

The cake you buy

During a single day a bakery may produce hundreds of cakes on a continuous production line.

As the materials are measured and mixed they are moved through the bakery on conveyor belts.

The designer of the cake production line has to ensure that all the stages will be correctly and rapidly carried out.

▲ *Mass production of cake in an industrial bakery.*

Fruit cake

The ingredients for a fruit cake may come from all over the world. When you bake a cake you buy and use small amounts. When an industrial bakery produces cakes it uses much larger quantities but follows the same process.

Production by an industrial bakery involves:
- buying and storing the ingredients;
- measuring out the required quantities;
- mixing the ingredients thoroughly and in the right order;
- baking.

The baked cake then has to be cut up, wrapped, packaged and delivered to the shops where you can buy it.

butter from New Zealand

cherries from Turkey

dried fruit from Israel

spices from the West Indies

wheat from Canada

eggs from the UK

dates from Egypt

▲ *Ingredients from around the world go into a fruit cake.*

Q 2 Draw a flow chart to show the stages in the industrial production of fruit cake. For each stage write down what has to be controlled so that the cakes produced are all of the same quality.

1 MEETING NEEDS AND WANTS

...Food and drink

Providing drinking water

We cannot live without water, so a readily available supply is essential. In many parts of the world people get their water from wells and water holes, and spend much time and effort in collecting and carrying it to where it is needed. Often it has to be boiled to make it safe to drink.

Mains water is moved from a reservoir by way of a water tower through pipes to people's homes. The water is held in a vessel at the top of the tower. This creates uniform pressure to deliver the water. It is filtered and treated with chlorine to make it fit to drink.

Q 1 What things could be used to make water easily available from a well? Explain the advantages and disadvantages of each in terms of:

a cost;
b complexity;
c ease of use;
d ease of maintenance.

▲ *Collecting water from a water hole.*

Different drinks

All over the world people have developed different ways to make drinks:

- Hot water is used to extract flavour from dried tea-leaves, herbs and ground roast coffee.
- Cold water is added to syrups to produce fruit squashes.
- Carbon dioxide gas is added to drinks to make them fizzy.
- A wide variety of fruits and grains are fermented or distilled to produce wines, beers and spirits.
- Yoghurt-based drinks have long been popular in the Middle East and Asia, and can now be found in our supermarkets. They may be sweet or salty and are very refreshing.

Q 2 List the different types of soft drinks in your local supermarket. For each drink write down what makes it appealing and for what occasion it would be suitable. List the ingredients for each and compare them.

▲ *A water tower.*

▲ *Some of the wide range of products from the soft drinks industry.*

Containers for drinks

Drinks are sold in a wide variety of containers including bottles, cans, cartons and plastic screw-top containers. These containers:

- prevent the drink from spoiling;
- stop the drink leaking out;
- are easy to open for drinking or pouring without spilling.

The designers of these containers try to make them easy to use:

- Some large plastic containers have a built-in handle.
- Plastic fizzy drink containers are shaped for easy gripping.
- Ring-pulls on canned drinks have been redesigned so that they do not fall off and become litter.
- Soft drink cartons have an easy-to-pierce region for use with a detachable straw.

The drinks containers used in fast-food outlets have to keep drinks hot or cold as well as allowing you to drink from them.

▲ *Drinks come in a wide range of containers.*

 3 Make quick sketches of different types of containers for drinks. Label the parts of the design that you think make them convenient to use.

Drinks from machines

Drinks can often be bought from automatic vending machines. The simplest of these allow you to choose from a selection of canned drinks.

Most machines give change. They release this into a small chamber for collection.

The more complex machines offer a selection of hot and cold drinks. A request for white coffee with sugar requires the machine to:

- select a cup containing powdered coffee, powdered milk and sugar;
- heat water and add it to the contents of the cup. Stirring is not necessary, because the water delivery system is designed so that the turbulence in the cup makes the powders and water mix evenly.

 4 Draw a flow chart describing the operations and decisions that an exact-money-only vending machine would have to make to sell you:

a a can of fizzy drink;
b a cup of black coffee with sugar.

How would the flow chart have to be altered for machines that give change?

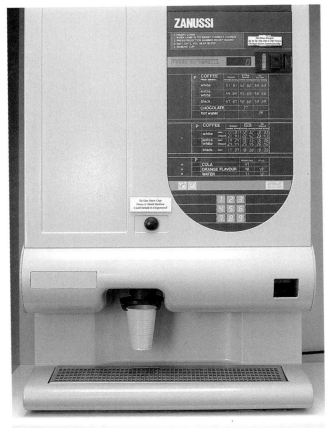
▲ *An automatic drinks vending machine.*

1 MEETING NEEDS AND WANTS

Carrying

People carrying things

Loads people carry include water from the well, tools and equipment, crops, firewood, goods to and from market and personal possessions.

 1 Different ways of carrying loads are shown in these pictures. For each one, decide:
- **a** which parts of the body are being used to lift the loads;
- **b** which parts are being used to support and balance the loads;
- **c** which parts may get tired or sore.

Designing carriers

People design carriers to make carrying easy. The design depends on:

- the load – how heavy it is, how bulky, how fragile, whether it is alive or dangerous;
- the people – how strong they are, their stamina, and what else they have to do whilst carrying;
- the method of carrying used;
- the materials available from which to make the carrier;
- how far the load has to be carried.

▲ *Four ways of carrying things.*

 2 Look at the carriers in the photos.
- **a** What is likely to be carried in each?
- **b** Is it heavy, bulky or fragile?
- **c** What materials are the carriers made from?
- **d** How will each carrier be used?
- **e** Which are suitable for long journeys?

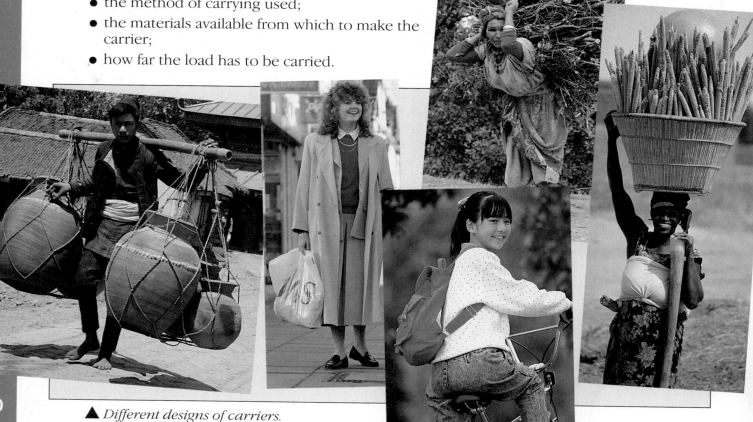

▲ *Different designs of carriers.*

Straps and handles on carriers must be comfortable and not cut into our bodies. They should be made from hardwearing, soft and pliable materials.

▲ *Straps and handles are important.*

 3 Look at a range of carriers to find out how comfortable they are to use. Record your findings.

Carrying babies

Babies need to be carried until they can walk on their own.
People have designed many ways of carrying babies. Babies are precious, so special attention is paid to their safety, warmth and comfort.

This baby is securely held by its father, but he will end up with aching arms and cannot do anything else at the same time. ▶

 4 Five ways of carrying babies are shown here.

a Which method would you choose for each of these age ranges: 0–3 months, 3–6 months, 6–18 months, 18 months–2 years?
b How does each method provide safety, warmth and comfort for the baby?
c How does each method make carrying the baby easier or more convenient?
d If you can, evaluate one or more of these methods by taking a user trip (see page 72).

Safety

Reducing the effect of road accidents

Designers have developed a range of products to make road travel safer. It is everyone's responsibility to make good use of these products.

Accidents to cyclists often cause head injuries. Wearing a safety helmet helps to protect the skull and brain.

Car seat belts prevent us from being thrown about in a crash or on sudden braking.

The front and rear of many cars are designed to crumple in a crash. This absorbs energy that might otherwise injure the occupants.

Lorries carrying harmful chemicals are labelled with 'Hazchem' warning signs so that any spills can be dealt with safely.

▲ *The warning sign tells us that this lorry is carrying highly flammable materials.*

Q 1 What material is used for making cycle helmets? Why has it been chosen?

2 Look closely at a car seat belt.
a What makes it so strong?
b What mechanism controls it so that it holds tightly only during an accident?

Preventing road accidents

Designers have developed systems and products to prevent road accidents from happening. Some are shown in these photographs.

Q 3 Explain how each of these helps to prevent accidents. What other systems or products help us to avoid road accidents? Explain how each of these works.

Product safety

Manufacturers must make sure that their products are safe for us to use. We are all responsible for using them carefully to make accidents unlikely. Both the design of products and the instructions for their use will affect their safety. Here are some examples.

Safe toys

Toys for the very young must be as safe as possible as children cannot see likely dangers:

- Non-toxic (non-harmful) paints must be used, as most small children put toys in their mouths.
- There must be no easily detachable small parts, as young children may swallow them or push them up their noses or into their ears.
- There must be no sharp edges or pointed parts with which small children could jab themselves.

▲ *These toys have been designed with safety in mind.*

 4 List the safety features of these toys.

Dangerous chemicals

Some household chemicals, like bleaches and disinfectants, are extremely toxic. The labels on the containers tell you of the risks, how to use the product safely and what to do in case of accidents. It is important that this information is clear and easy to understand.

 5 Does the label on the container give all the information needed to use the product safely? What other design features help to ensure that the products are used safely?

You might find a label like this on a bottle of bleach. ▶

Leads can kill

People may be badly scalded when the lead to an electric kettle catches on something and the kettle tips up.

The cordless kettle overcomes this hazard as the lead is connected to the base plate. The kettle can be plugged directly into the base plate to receive electrical power and then lifted free from it for pouring or filling.

▲ *The cordless kettle prevents problems with leads.*

 6 For each of the following products write instructions that will help people to use the product safely. Give clear warnings about dangers:

a hair drier;
b solvent adhesive;
c frozen food;
d winter pyjamas.

1 MEETING NEEDS AND WANTS

...Safety

Warnings

Designers have produced warning devices for our safety and protection. These devices must:
- sense the danger;
- operate an alarm.

The alarm is usually an unusual and loud noise, sometimes with a flashing light. The type of **sensor** will depend on the danger to be detected. Here are two examples.

Intruder alarms

Here the warning device must detect the presence of someone trying to break into a building. A simple form uses sensors called reed switches, which can detect movement. These are placed on doors and windows and sense when they are opened. Then the alarm sounds. Sometimes these **systems** are connected to the local police station.

The control panel and alarm box of a modern intruder alarm.

More complex systems use light beams to detect intruders. If the light beam is interrupted, the alarm sounds. Even more elaborate systems use changes in the frequency of ultrasonic sound to detect the movement of an intruder inside a building.

It is important that an alarm is designed so that it cannot be 'turned off' from outside the building it is protecting.

Vehicle safety warnings

Just behind a large lorry there is a point that the driver cannot see even with the aid of wing mirrors. It is all too easy for someone to walk behind the lorry as it starts to reverse, and they may be seriously injured. To avoid this, many large lorries are fitted with reversing alarms which sound automatically when the lorry is put into reverse gear.

▲ *This person is in the danger zone.*

Q 1 List places where warning systems might be useful. List the sensors you would need in each case and describe how the alarm would work.

Recreation

Young or old, we all need recreation. The child pretending to be a pirate, the teenager learning karate, the family on a day out at the seaside and the elderly couple playing bowls are all examples of how we use recreation to refresh our minds and bodies. This section shows some ways to use design and technology to provide people with recreation.

▲ *Some of the board games on sale today.*

Board games

A wide variety of board games is available. Some, such as chess and draughts, have been played for centuries. Others are more recent. Fantasy adventure games like 'Dungeons and Dragons' cater for particular enthusiasms. Quiz games such as 'Trivial Pursuit' are designed for general family entertainment. Games like ludo and snakes and ladders rely entirely on chance and help children learn to count.

Most board games have the following parts:

- a board on which the game is played;
- a set of counters or pieces that are moved around the board;
- spinners or throwing dice if there is a chance element;
- rules for playing the game;
- sets of information and penalty/advantage cards.
- a box with a cover indicating how exciting, enjoyable or complicated the game is.

To develop a new board game you should design all these parts.

▲ *All the parts of a board game must be designed.*

Q 2 List the board games you have enjoyed playing, and for each one explain why you enjoyed it. Use your list of comments to explain why board games are popular.

1 MEETING NEEDS AND WANTS

...Recreation

Puppet theatre

Most children enjoy puppet theatre performances, and in many countries they are enjoyed by everyone. Performances usually involve telling a story using 'larger-than-life' characters. The characters' features are exaggerated so that the personality is obvious. Often the story has a special meaning for those watching. The characters may act out important events in the people's history. The story may be used to tell about the values of the society. It may be a story telling about their beliefs.

In producing a puppet theatre and performance you must take into account:

- **The audience** – Make sure that they know when and where the performance is and that the story is suitable for them. Organize the seating so that the performance can be seen clearly from all seats.
- **The story** – Choose an existing one or write your own. A clear story-line with only a few characters is usually best.
- **The puppets** – Decide what type of puppets to use and how to make them 'work'.
- **The characters** – Design both faces and costumes. Make sure that these match the characters and can easily be seen.
- **The theatre** – This can be very simple, perhaps a curtain on a string between two tables. Or it can be specially constructed with backdrops, lighting and curtains which can be opened by pulling on a rope. There must be enough room in the performance area for the puppets and the puppeteers. Puppeteers can wear black costumes and operate the puppets against black curtains. Then no theatre is required and the performance area can be larger.
- **The performance** – This will be a team effort and you will need to practise before the real performance.

Q 1 Describe how each of these puppets is constructed.
How does the operator 'work' each type of puppet?

▲ *Puppets come in all shapes and sizes with different ways of working.*

Toys

Children learn while they are playing. Toy manufacturers respond to children's needs by producing a wide range of playthings. Those who design toys should be aware of what a child might learn through playing with a particular toy.

The toys in the picture are small-scale models for children aged 5 to 12 years. Children playing with them would act out a story where the toys were the main characters. Examples of stories to act out appear in comics and television cartoons about the toy characters. Children playing with the animal might learn to be gentle and to care for others. Children playing with the robot might learn to be fierce and ruthless towards anyone who upsets them.

 2 List some of the toys you have played with over the past three years. For each one write down what you might have learned through playing with it.

▲ *Two popular toys linked with television and comics.* ▼

Playgrounds

Playing in the street or on waste ground can be dangerous. The wide open spaces of parks can be boring. For places to play that are safe yet exciting designers have made playgrounds. They contain two main sorts of play equipment:

- **Structures** – You can climb over and through these while having imaginary adventures and games. Being up high or hidden away adds to the excitement.
- **Rides** – These include swings, roundabouts, slides, seesaws and rocking horses. The movement and the thrill of taking risks provide the excitement.

The surface of the playground is important. If it is too hard a fall may cause injury. If it is too spongy it is not suitable for running.

Evaluate your local playground like this:

- Draw a plan showing the overall layout and the play equipment.
- Make sketches or take photographs of each piece of equipment and put them in a scrapbook.
- For each piece of equipment add comments on how exciting and how safe it is.
- Say whether the playground as a whole is safe and attractive.

 3 Look at the playground equipment in the picture. What dangers can you see for both users and passers-by?

 Playgrounds should be safe and exciting.

Interior design

Things to consider

An interior designer's job is to enable us to live and work in places that are comfortable, attractive and functional. The designer of this office had to think about:

- what people use the room and what they need to do;
- the shape and size of the room;
- the availability of gas, electricity and water;
- the positions of doors and windows;
- the furniture and how to arrange it;
- any special equipment and where to put it;
- the colour scheme and lighting;
- furnishings that could add comfort and decoration;
- final touches to make it attractive;
- any legal requirements.

▲ *Office design should ensure that everyone can do their job in comfort.*

Q 1 Would you like to work in this office? What do you find attractive or unattractive about it?

Tackling the task

Using plans

Interior designers use plans to help them make design decisions. A basic plan shows the position of the walls, doors, windows and electricity, gas and water points. The plan can also show which way the room is facing and how much natural light it will receive.

Card cut-outs will help the designer try out different arrangements of furniture and equipment and think about the advantages and disadvantages of each layout. The layouts will also show how people and materials might move around or through the room.

Using drawings

To help decide on the colour scheme and furnishings the interior designer draws a perspective view of the room and tries out different colours and decoration schemes. She can add samples of fabric and wallpaper.

Q 2 Draw a plan of a room at home. Mark in what you would change.

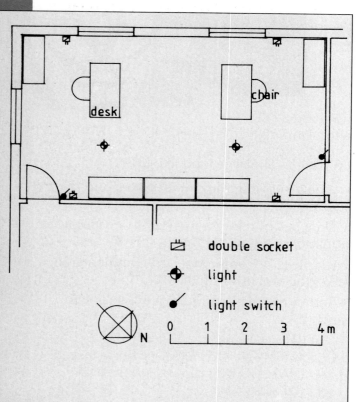

▲ *An office plan.*

Three interiors

The kitchen-diner

The family who use this kitchen-diner want to be able to:

- prepare and cook food and drink;
- serve and eat meals;
- clear away and keep the place clean and tidy;
- make it comfortable and attractive.

Meeting all these needs in a small area will challenge the designer.

▲ *A kitchen-diner for a small family.*

The doctor's waiting room

People who use a doctor's waiting room are often worried or nervous, so it is important that the design of the room, its furniture and fittings, put them at their ease. A communication system is needed to call patients when it is their turn to see the doctor. Patients with small children may need special facilities.

▲ *A doctor's waiting room.*

 3 How do the interiors in these two pictures meet the needs of the users? Would you change anything?

John's bedroom

John's hobby was subaqua, so he used a fantasy underwater-world theme to decorate his bedroom. He printed the designs on the bedcover and curtains and painted the shipwreck scene on the wall.

 4 What theme would you choose for your bedroom? What images would you use to decorate the walls and furnishings?

▲ *A fantasy theme for a bedroom.*

2 Strategies

Working methods for designing

Each design and make task you do will be made up of lots of smaller tasks. You will need to use different methods or **strategies** for each of these. This chapter describes some of these strategies.

When you are tackling a small task try working like this:

- Ask yourself, 'What ways are there of doing this small task?' (There is usually more than one way.)
- Use this chapter to find out about some possible strategies.
- Choose the strategy that looks best but keep your options open. If it does not seem to be working try one of the others.

Your teacher may help you to learn some of these strategies by setting you resource tasks that give you practice.

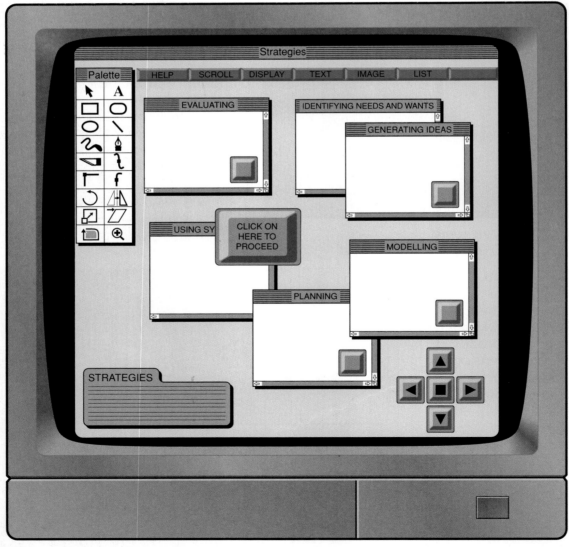

▲ For success in tackling a task you need to choose the right strategy.

Identifying needs and likes

What to look for

In designing something it is important to find out the users' needs and what they like (their preferences).

Look at these three important areas:

- **The place** – Is it small or large, crowded or empty, noisy or quiet, well-lit or gloomy, pleasant, comfortable?

- **The people** – Is there a wide range of ages, cultures, sizes and appearance and different genders, or are they all similar? What are they doing? Are they having difficulties? What can you tell from their faces?

- **The existing products and systems** – Are they easy to use? Do they work well? Could they be improved?

▲ *There are lots of needs and preferences here.*

Tools for recording

You can record your findings in many different ways:

Notes – You only need a note pad and a pencil. Write up your notes in more detail soon afterwards.

Sketches – A quick sketch is often better than a lot of writing. Write short notes on your sketch to explain the important points.

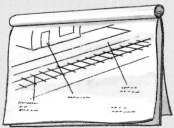

Audiotapes – Before you start using a pocket tape recorder check that it works, is adjusted to your voice level and that other noises do not drown out your voice. Carry spare batteries and tapes. Producing a transcript (written copy) can take a long time.

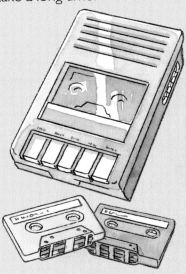

Photographs – This is a good way to record a situation. First you need to learn how to use the camera. Keep brief notes on each shot you take and use them as captions for the photographs.

Videotapes – The equipment for video recording is expensive and you will need training to use it. Many people behave differently if they realize they are being filmed, so you have to be discreet. The tape may need editing, with a commentary. This can take a long time.

Resource task

SRT 1

...Identifying needs and likes

Interviewing

Sometimes the only way to find out about something is to ask someone who knows. You will need to conduct an interview. Here is some useful advice:

- Dress tidily and be polite.
- Ask for permission if you want to tape record the interview.
- Have a list of questions ready plus a pencil and pad to jot down answers.
- Some people like to see the questions beforehand to prepare their answers.
- Do all you can to put the person at ease. Smile, be friendly and, if possible, hold the interview in a room where you can both sit down and feel comfortable.
- Make sure that the person understands the question and give her the chance to ask if she is not sure.
- Say 'Thank you' at the end of the interview.

▲ *Conducting an interview.*

Looking in books and magazines

Other people may already have written about your subject in books or magazines.

In libraries books are organized into subjects using the Dewey decimal numbering system. You can look up the numbers for a subject. The books will be arranged on the shelves according to these numbers.

You may have to choose between several different books. Use the contents pages to find out what each is about, and read the introduction to find out its level. You may be able to borrow it. If it is for reference only, you can make notes or take photocopies of important pages.

Specialist magazines, on sale at large newsagents, report on the latest developments in technology and the technical aspects of sports or hobbies.

▲ *The choice of information sources can sometimes be bewildering!*

Resource task

SRT 2

Analysing needs

Designers and technologists must understand people's needs so as to be able to meet them. The **PIES** checklist helps us to think about needs. Each letter stands for a different type of need.

P We all need food, water and air to breathe. We need to keep warm and be protected from the weather. We need regular exercise. These are **physical** needs.

I We need to learn new things and to be stimulated. We use games, books, television and radio, and so on, to meet these **intellectual** needs.

E We all need to feel safe. We need to feel that people care about us, and to have ways of expressing our feelings. These are **emotional** needs.

S Most of us like to spend time with our friends, talking and doing things together. These are **social** needs.

▲ *The PIES checklist will help you identify these people's needs.*

Exploring style

People of different cultures have different views on what is visually pleasing, and even within the same culture people often disagree. Fashion and changing style influence what people like. It is important to be aware of these differences when you are designing. The area of design concerned with how a product looks is called **aesthetics**.

These pieces of furniture come from a variety of times and cultures. ▶

Resource tasks

SRT 3, 6

...Identifying needs and likes

Looking at colour

Colour is important in our lives. It can make us feel excitement and happiness or relaxation and sadness. Designers use this knowledge when designing products.

You can set up colour preference tests (using colour cards and a questionnaire) quite easily to give you an idea of how people react to certain colours. But you also need to think about how and where the colour is to be used: even if blue is someone's favourite colour, they might not be too happy if you gave them blue food!

▲ *Red gives this car an exciting appearance.*

◄ *A colour wheel shows a complete range of colours to help you make choices.*

Colour is important in helping us to identify the flavour of food. We can easily be fooled when the familiar colours of foods are altered or disguised. For example, people find it difficult to identify the flavours of soft drinks when they are blindfolded.

Colour influences people's behaviour. Red can make us feel warm and comfortable, but can also make us feel restless after a while.

▲ *Colour often gives a clue about the flavour of a drink.*

▲ *A colour scheme may encourage you to remain seated or to get up once the meal is over.*

Resource task

SRT 7

Advertisers and manufacturers use colour to shock or gain our attention. They have also discovered that we associate certain colours with certain types of product.

You would expect to find different types of product in different colour packages. ▶

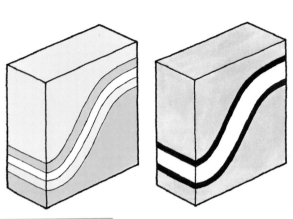

Image boards

One way of thinking about the people for whom you are designing is to set up an image board. This is a collection of pictures – cut-outs from magazines or newspapers, quick sketches, specially taken photographs – to do with the people. Pictures of the people themselves, where they live, where they work or go to school, what they do in their leisure time, the food they eat, the clothes they wear, where they shop – these can all go on the image board, which will help you understand the people's life-style and what might appeal to them.

Resource tasks

SRT 4, 5

Design briefs

Understanding a design brief

When you are designing, you will start by exploring a situation where there is a need or want. If you decide that there is a practical problem that you can solve, you will need a **design brief**. This summarizes the aim of a design project and states briefly the kind of thing that is needed. Here is an example.

The messy desk top in the picture above presents an opportunity for design. There is a need for some way of keeping the desk top tidy. The design brief might be:

> ❛Find a way to keep drawing equipment tidy on a desk top.❜

There are many possible solutions to this design brief. The picture on the right shows some of them, ranging from simple containers to more complicated racking systems, and even a computer-aided drawing system. This is an **open brief** and allows the designer a large degree of freedom to experiment with design proposals. An open brief does not say what the solution to the problem is going to be.

▲ *Ways of keeping a desk tidy.*

Sometimes a design brief narrows the likely outcome. For example, the design brief for the same situation as above could have been:

> ❛Design and make a pencil box.❜

Although it is still possible to produce a variety of different solutions from the brief, the kind of product that is to be made is more limited. This is a **closed brief**.

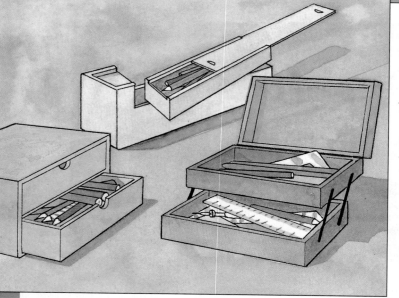

▲ *Different designs for a pencil box.*

Writing a design brief

When writing a brief, choose the words that you use carefully. It is best not to make your design brief too detailed. For example, instead of writing the brief 'Design and make a pencil box', use a more general term like 'container' or 'desk tidy'. This offers you a much wider choice of solutions, and you may find a more interesting way of solving the problem of clutter on a desk than making a pencil box!

Other good words to use in design briefs are 'device', 'item', 'system' and 'could'. For example:

- Design and make a device to help a young child learn to tell the time.
- Design and make a storage system that could be used in a nursery.
- Design and make a container for snack food.

Here are some situations for design, and design briefs that have been written in response to them.

DESIGN BRIEF
Design and make items that encourage safe, exciting and imaginative play in a playgroup.

DESIGN BRIEF
Design and make a range of snack foods that could be eaten by young children at school.

DESIGN BRIEF
Design and make a range of items to enable messy activities, such as painting, to be carried out in a nursery.

Resource tasks
SRT 8, 9

Specifying the product

Writing a performance specification

On starting a design and make task you will need to develop the design brief into a **performance specification**.

This should always:

- describe what the product has to do.

It might also state:

- what the product should look like;
- that the product should work in a particular way – use a particular material or energy source, perhaps;
- any legal or environmental requirements the product should meet.

You can use a specification to check your design ideas as they develop. In this way you avoid designing something that does not meet the requirements. You will also need to check the finished product against it.

Examples of some products and their specifications are shown on these two pages.

SPECIFICATION
SCHOOL EQUIPMENT BAG

What it has to do:
- hold pens, pencils, rubber, ruler and stencils
- stay shut when closed but be easy to open
- be tough enough to withstand everyday use
- be small enough to go in a rucksack
- be easy to carry about on its own

What it has to look like:
- be brightly coloured
- have a modern look and indicate who it belongs to

Other requirements:
- use biodegradable materials as far as possible

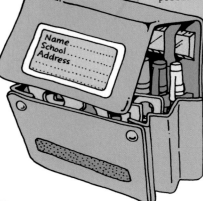

SPECIFICATION
PIECE OF JEWELLERY

What it has to do:
- be suitable for everyday wear
- remind people about endangered species

What it has to look like:
- be based on animal or plant form

Other requirements:
- be part of a range of similar items for teenagers sold in a chain store
- use natural materials as far as possible

Other considerations

You will need to take into account other things as you develop your design ideas. You should ask yourself:

- How much time do I have for designing and making?
- What tools and equipment are available?
- What materials and components are available?
- How much money do I have to spend?

Using a performance specification

Once you have some design ideas, review them against the specification. Ask:

- Will it do the job?
- Will it look right?
- Will it work?
- Is it a practical suggestion?

Answering such questions will ensure that your designing gets off to a good start.

As you develop your design ideas, it is important to ask yourself these questions again, to avoid losing sight of what the product has to do.

Once you have made your product you need to evaluate it. Evaluating it against the performance specification is one way to do this but there are other important evaluation techniques (see pages 72–75).

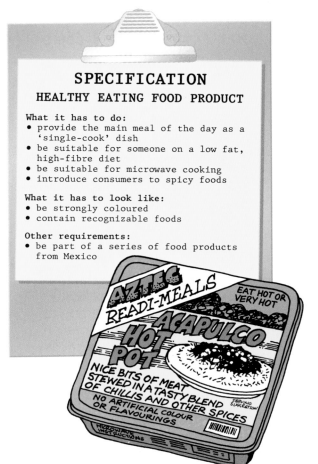

SPECIFICATION
HEALTHY EATING FOOD PRODUCT

What it has to do:
- provide the main meal of the day as a 'single-cook' dish
- be suitable for someone on a low fat, high-fibre diet
- be suitable for microwave cooking
- introduce consumers to spicy foods

What it has to look like:
- be strongly coloured
- contain recognizable foods

Other requirements:
- be part of a series of food products from Mexico

SPECIFICATION
INTRUDER ALARM

What it has to do:
- detect break-ins through doors or windows
- sound a loud alarm
- show on a display panel where the break-in happened

What it has to look like:
- its presence should be obvious and not hidden
- its visible parts should look modern
- the display panel should be easy to read

Other requirements:
- it should be impossible to deactivate it from outside the building

Revising the specification

Sometimes you will have an exciting design idea which does not meet the specification. It is very tempting to change the specification so that it fits the new idea. This can lead to serious problems unless everyone involved with the design is consulted and agrees to the changes.

Resource task
SRT 10

Generating design ideas

Brainstorming

Brainstorming is a good way for a group of people to generate (create) lots of ideas quickly. This is how to do it.

1. **State the problem or need.** Write down the problem on a large sheet of paper or a chalkboard. Be sure to record the problem and not what you think the solution might be.
2. **Write down every idea suggested.** One member of the group can write down everything, even if the ideas or words seem silly at first. Use words, phrases or pictures to capture the ideas. As suggestions are recorded, they will spark off more ideas.
3. **Concentrate on quantity.** Produce as many ideas as possible. This gives you lots to choose from when it comes to picking the best.

4. **Don't make judgements.** If you say someone's idea is stupid, it may stop them producing more.
5. **Allow a set time for the session.** The first part of a brainstorming session usually produces the more obvious ideas and thoughts, so allow enough time for some unusual ideas to emerge.
6. **Sorting out the ideas.** The ideas collected now have to be sorted out. Repeated ideas should be removed. Some will be impossible to use: they might be too difficult, cost too much or take too much time. This will leave a set of new ideas that can be used as starting points for design and technology.

▲ Sorting out the ideas.

Knowing when and how to use brainstorming

Brainstorming does not work for problems that have only one solution, such as the size of the population of France or the sales of television sets. You would look these up in a book or on a database. Brainstorming does not work well if one member of the group knows a lot more than the others about the subject of the problem.

Brainstorming only works if:
- people don't butt in;
- the ideas are short;
- people don't criticize one another's ideas;
- the ideas are imaginative;
- the group members use the earlier ideas to spark off more ideas.

Bubble charts

Before you can start work on any design brief you need to ask: *what? where? when? who?* and *why?* The answers will give you a better understanding of the problem. It is also a good way of generating ideas for design solutions. One way of recording your answers to these questions is to draw a bubble chart.

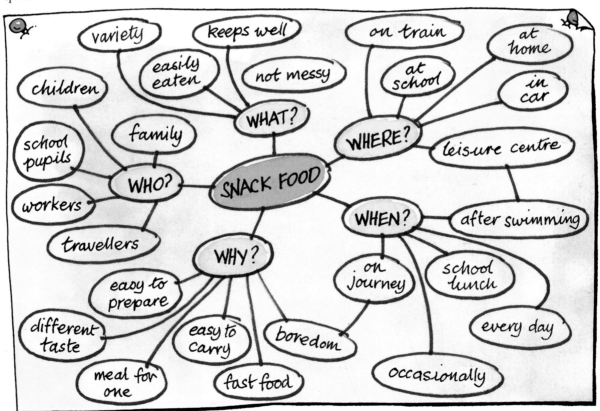

▲ *A bubble chart helps you to understand the problem.*

Start by writing a key word or even a design brief statement in a bubble in the centre of the page. Then use the questions to generate more words and ideas. You can combine this with brainstorming to come up with ideas.

Resource task

SRT 11

... Generating design ideas

Observational drawing

The world around us has been a rich source of ideas and inspiration for many designers and inventors. Leonardo da Vinci made sketches of subjects from nature which inspired his wonderful ideas and inventions.

The example below shows how close study through drawing has resulted in designs for jewellery with a pattern.

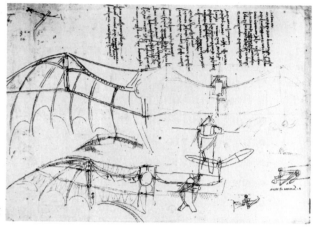

Leonardo da Vinci developed designs for aircraft wings by observing and drawing birds' wings. ▶

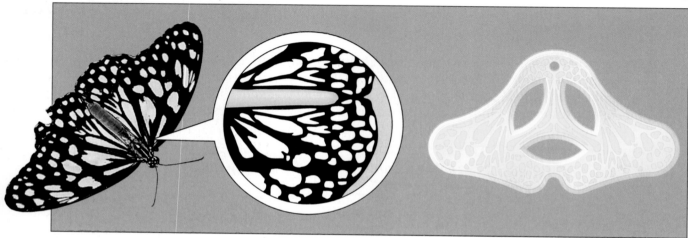

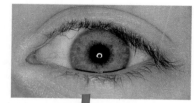

▲ *This brooch design is based on a butterfly although it doesn't look like one.*

Making connections

Design ideas are rarely completely new. Often the idea may have been suggested by something in nature, or by connecting two different ideas in a new way. For example, if you needed to provide portable shelter for a hill-walker, you might look at natural or man-made forms of protection: umbrella, snail shell, bus shelter, seed pod, eyelid, dustbin liner. Each of these suggests a different way of solving the problem for the hill-walker. Each provides an interesting starting point for further development.

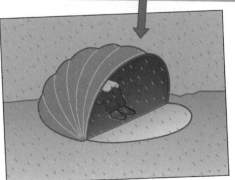

Resource tasks
SRT 12, 13, 14, 15

▲ *Connecting ideas can help generate new designs.*

Attribute analysis

Designers and technologists use a technique called **attribute analysis** to help them produce new designs for familiar objects.

A group of pupils used attribute analysis to design a pen which a garden centre might give away to visitors. This is how they did it:

1. They wrote down all the different words they could think of to describe the pen.
2. Then they looked at each word to see what it represented, as in the photograph.
3. They used these ideas as headings for a table like this:

The body of the pen is made from plastic – a **material**.
The pen is a tube – a **shape**.

The pen is rigid and this is to do with the material's **properties**.

It is light and this is to do with its **weight**.

It is cheap and this describes **cost**.

Material	Shape	Properties	Weight	Cost

4. As a group, the pupils brainstormed entries for each column. In the 'material' column, for instance, they wrote down plastic, metal, glass, pottery, wood and stone. They kept the idea of a garden centre in mind to suggest some ideas. This is their completed table:

An attribute analysis table for producing new designs of pen. ▼

Material	Shape	Properties	Weight	Cost
plastic	tube	rigid	light	cheap
metal	cube	soft	heavy	expensive
glass	leaf	flexible		
pottery	leaf			
wood	spiral			
stone				

5. By looking across their table in different ways they produced different ideas. One was for a metal, spiral-shaped, flexible, heavy, expensive pen. They decided to make a wooden, leaf-shaped pen that was rigid, light and cheap as the gift for the garden centre. What would you decide?

You can use attribute analysis to help you think about the design of any product.

Resource task
SRT 16

2 STRATEGIES

Modelling – how it can help

When you come up with a design idea, you are the only one who knows anything about it. While it is just an idea you cannot test it, see what it will look like or know that it will work. Modelling helps you to convert the idea into a form that you can think about more easily and show to other people to get their opinions.

Sometimes you can model your idea by talking about it. At other times you will need to model it on paper using sketches and notes. Often, as you do this modelling you will realize that there are things about your idea that are not clear enough yet. So it has to be rethought out for you to come up with a better description or clearer sketch.

You may need to work in three dimensions to get a really clear picture of your design.

Some models will help you to see how it will look, while others will help you to see how it might work.

You can use computers to model your design ideas too. For example, some software can help with electronic circuit design.

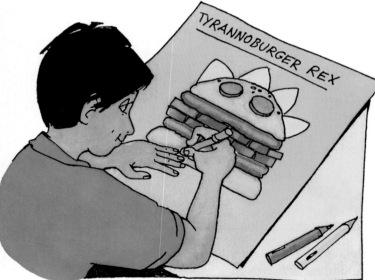

Pages 45–65 describe the different methods of modelling that you can use to help you when you are designing.

You can use modelling in three ways:

- to clarify and develop your design ideas;
- to evaluate your design ideas;
- to show your design ideas to other people.

Modelling product appearance on paper

Sketches as well as talking help define design ideas. Most designers model their ideas by sketching them, adding notes as they go to make things clearer.

These quick drawings are called **annotated sketches**. Although they are rough, they are very useful to you and other members of the design team because they help you think about your ideas.

Simple shapes and guidelines

Use faint pencil guidelines when drawing simple shapes.

If the object you are sketching is symmetrical (both halves the same), a lightly drawn centre-line can act as a useful guide. For irregular shapes, mark out the main features so that you can position everything correctly.

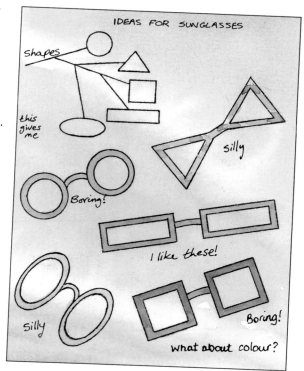

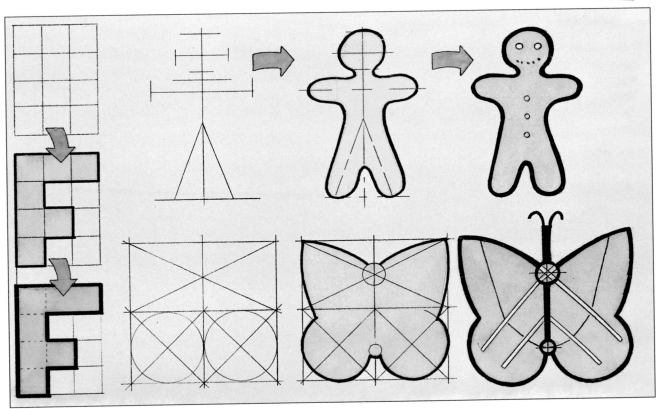

Keep all your sketched lines faint until you are sure that you have drawn the correct shape. Then go over the outlines to make the shape stand out from any guidelines.

Resource task
SRT 17

...Modelling product appearance on paper

Grids

Grid papers are useful for mapping or planning arrangements of shapes. You can sketch garden plans, room layouts and circuit boards more easily and accurately on squared grid paper.

Grids are also good for developing interesting patterns. You can plot more complex shapes, and use different types of grid paper to experiment with pattern. With a grid, you can repeat a pattern accurately and regularly.

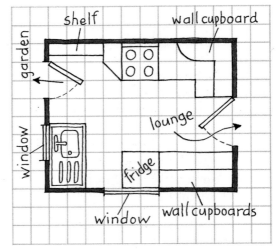
▲ *Planning a room layout on grid paper.*

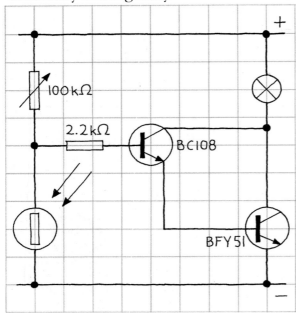
▲ *A circuit board layout.*

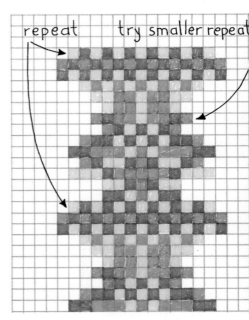
▲ *Using a grid to experiment with repeat patterns.*

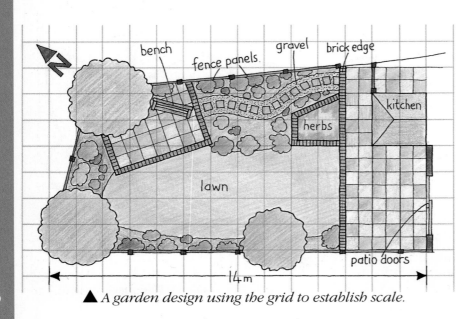
▲ *A garden design using the grid to establish scale.*

Resource tasks
SRT 18, 19

Quick 3D views

To draw a three-dimensional (3D) view of something start with a flat, two-dimensional (2D) shape – say, a front or side view of the object. Draw in parallel lines from the corners or edges of the shape at an angle of about 45°. Joining up these lines will give you a 3D form.

This view of an object, an **oblique view**, is an easy way of drawing circular or curved forms because oblique views are always based on 'flat' front or side views.

You can use the squares on grid paper as a guide when drawing the basic shape and the parallel lines. If you don't want the grid lines to appear on your finished sketch, draw on a piece of thin paper laid over the grid and fixed down with tape or clips.

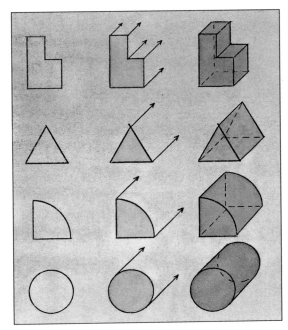

Simple shapes made 3D. ▶

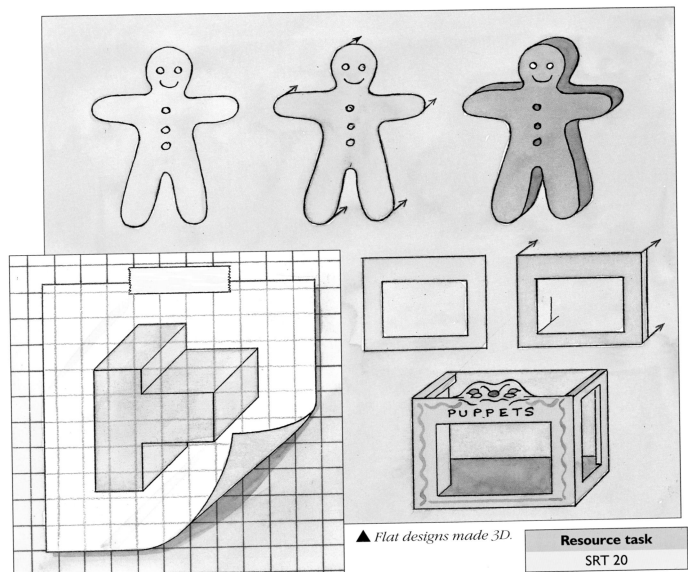

▲ *Flat designs made 3D.*

Resource task

SRT 20

...Modelling product appearance on paper

Crating

This is useful for drawing more complicated objects. Imagine that the object is in a box or crate, and use that as a starting point for your sketch.

Draw the box as a 3D view (see page 47), then draw your object by taking parts away from the box.

You may need to break the object down into several boxes to create the shape you want.

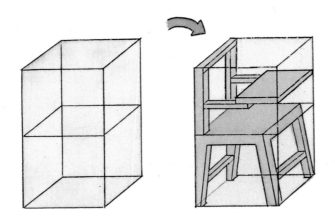
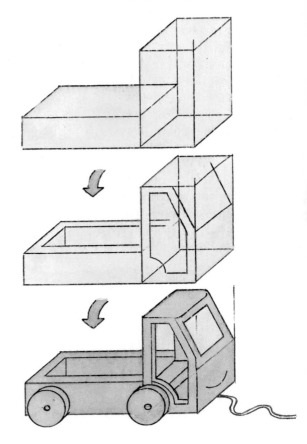

▲ *Using crating for 3D sketches.*

Thick and thin lines

When you are using guidelines and crating, it is important to make your final outlines stand out from other lines on the paper. A darker outline distinguishes the shape from the construction lines.

Vary line thickness to help make the object you have drawn look more solid and three-dimensional. But do not overdo this. Use the simple rules in the picture as a guide.

1 Add a thick line to an edge where only one surface is visible.
2 Leave a thin line where two visible surfaces meet.
3 For extra impact, draw in the thickest line around the outline of the object.

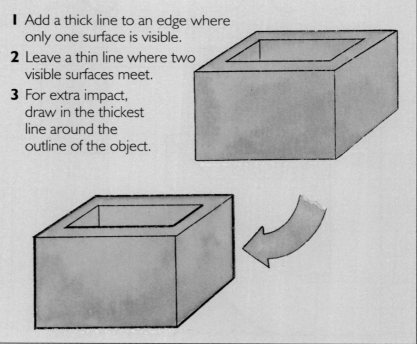

▲ *Using thick and thin lines.*

Resource task
SRT 21

2 STRATEGIES

Making things look solid

Line drawings can describe flat shapes and 3D forms, but lines only show a framework. In reality, we cannot see these lines on objects.

Instead, we notice that one surface is darker than another because of the effects of light and shade.

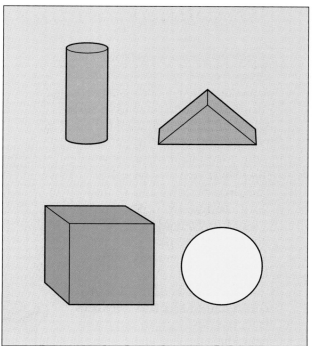

Shading makes the object you have drawn look more solid. It shows the effect of light falling on its surface.

The variations of light and dark are called **tone**.

The simplest way to show difference in tone is to use pencil or pencil-crayon shading. This will go from very light, for the side of the object closest to the light source, to very dark, for the part that is in shade.

▼ *Shading can make objects look solid.*

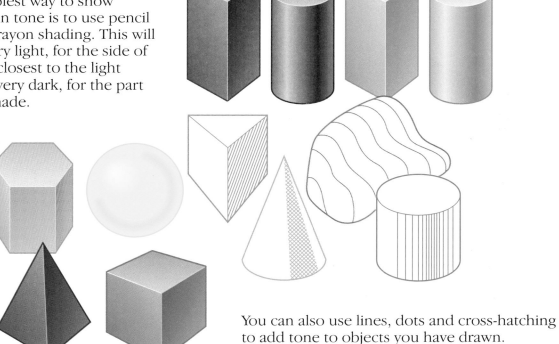

You can also use lines, dots and cross-hatching to add tone to objects you have drawn.

▲ *Some different methods of adding tone to your drawings.*

Resource task
SRT 22

49

2 STRATEGIES

Modelling product appearance in 3D

Rapid model making

Making a 3D model, using quick techniques and easy-to-work materials, is a useful way of finding out whether your ideas look right. It may also help you see whether the proportions and scale of your design are right.

Seeing your idea in 3D at an early stage may help to avoid expensive mistakes later on.

These kinds of models are often called **sketch models** because you can make them quickly.

Trying out ideas for a piece of jewellery using paper taped on to wire. ▶

Key points:
- Decide on the level of detail you need. (Remember, these are only intended to give quick impressions.)
- Use the most appropriate materials for the model. (For example, you would not use Plasticine to see what a pop-up card would look like!)
- Choose a suitable scale for the model.
- Choose materials that are easy to work or assemble.

Using foam models to help make a decision about scale. ▶

Materials for rapid model making

Plastic foams
Foam is light but rigid and good for making block models. It is easy to cut with any blade and can be shaped with a file or rasp. **(Beware of dust particles – use a mask.)**

A hot-wire cutter is effective for shaping foam blocks, but can result in dangerous fumes, so, again, great care is needed. **The heating wire should never become red-hot. Keep a window or door open to prevent fumes building up.**

To join pieces together, use a special adhesive – the wrong sort will attack the plastic foam. Temporary joints can be made with cocktail sticks.

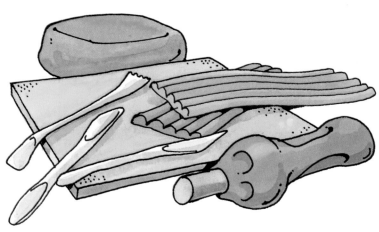

Found materials: packaging, fabrics
Cans, bottles, polystyrene foam, card and cardboard tubes are useful sources of shapes and forms for experimenting.

Reclaimed fabrics or remnants can be used for ideas for textiles or to make sketch models of fashion items.

Paper, newsprint, wrapping paper, wallpaper
These are suitable for rapid modelling of ideas involving textiles or where patterns and templates might be used. They are also good for quick packaging models.

Work quickly: cut, tear and fold. Fix using a glue stick, stapler or tape. Double-sided tape is particularly useful.

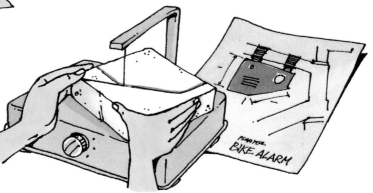

Plasticine and clay
These are easy to mould into shape, and so are good materials for modelling a small, irregular object. Details and texture can be added by pressing onto the surface with cocktail sticks, pencils or textured boards.

Plasticine can be reused. Clay is easy to mould, but can be messy and will set hard.

Resource task
SRT 23

2 STRATEGIES

Modelling product performance on paper

Sketches and diagrams can help you to see how your design might work, as well as look.

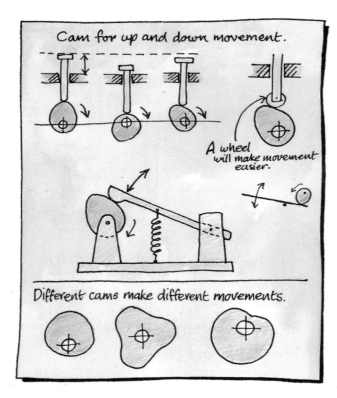

▲ Thumbnail sketches of how a design might work. ▼

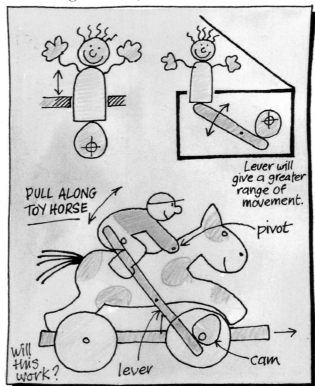

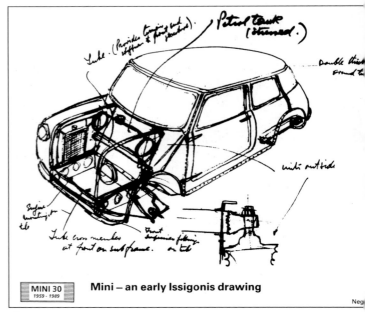

Mini – an early Issigonis drawing

▲ A thumbnail sketch by Issigonis, the designer of the Mini.

Thumbnail sketches

As you begin planning your design, it is enough to sketch the main parts.

These rapid sketches are called **thumbnails**. They should show very little detail and help you sort out your thinking on paper. Because thumbnails allow you to put ideas down quickly, they do not slow down the flow of your design thinking.

Remember to:

- keep the drawings or diagrams quick and simple;
- use a pencil, fine line marker or biro – whatever you find easiest;
- draw flat, 2D outlines without taking time to shade and colour;
- use only stick figures for people;
- use a word or notes to make things clearer;
- use symbols such as arrows to show movement or direction.

Try out any 'shorthand' techniques that you can think of to help you put down *only* the essential information.

As you become more practised at quick sketching, you will find that the drawings can actually suggest new ideas (see page 42).

Maps and plans

As your design idea becomes clearer, you may need to make more accurate 2D drawings that show greater detail.

Maps and plans can show the location of objects or components. A plan is a view of an object or area seen from above. It is sometimes called a bird's-eye view. It may be used to show the position of objects in a room, or the layout of smaller details, such as the buttons on a remote control unit or components on a circuit board (see page 46).

A map can be useful for plotting the movement of people or traffic, or for locating places at a distance.

▲ *Using a plan to arrange kitchen units.*

Guidelines for clearer maps and plans

- Use different colours or symbols to make information easier to understand. For example, you could plot the routes of different people around a room using different colours to avoid confusion.
- Use symbols to indicate objects.
- Use dotted lines or arrows to show movement.
- Always include a key to explain your symbols.
- If possible, include an indication of scale or size.

▲ *The layout of an easy-to-use remote control unit.*

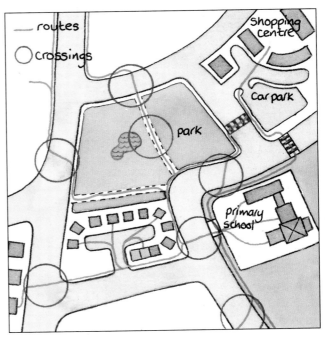

▲ *Plotting routes from school.*

...Modelling product performance on paper

Sometimes you will need to draw a 3D view of your design to see how it might fit together and work. These two pages describe some techniques that you could use.

Hidden details

Often the working details of an object are hidden inside. You can show these hidden details on a drawing in several ways:

- Draw dotted lines to show the outline of the shapes hidden inside the object.
- Make a **see-through drawing** of your design.
- Draw what you would see if you cut through the object – a **section drawing** of your design. This can show the thickness of the construction material.

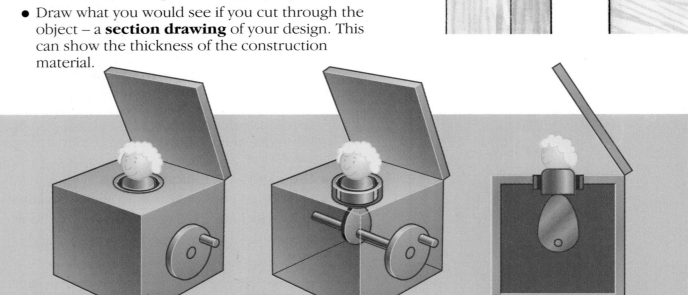

normal view　　　'see through' view　　　sectional view

Magnified details

The smaller details on a drawing may be difficult to see or understand. You can overcome this problem by imagining that you are holding a magnifying glass over the detailed part and draw an enlarged view of it.

Drawing a frame around the magnified detail helps to focus attention on it.

Use this method to show construction details that would not normally be seen.

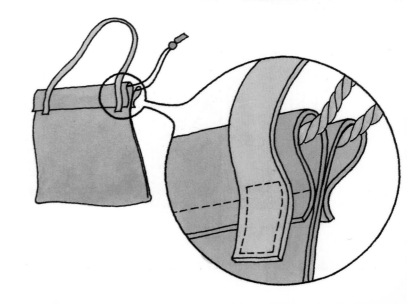

Cut-away views

The cut-away view allows you to show the construction and internal details of your design. Layers or parts of the object are removed or **cut away** to enable these details to be seen.

A cut-away view should only be a sketch. There is no point in producing a highly finished illustration.

Exploded views

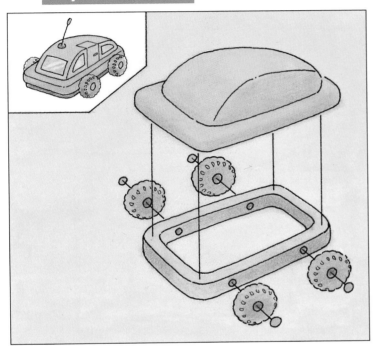

An **exploded view** is a useful way of showing how all the parts of a design idea fit together.

The title suggests an explosion, but the drawing really shows the object as if it had been pulled apart.

The pieces should be organized to show how they fit together. Arrange the parts along straight lines and put faint lines in to show the connections, if necessary.

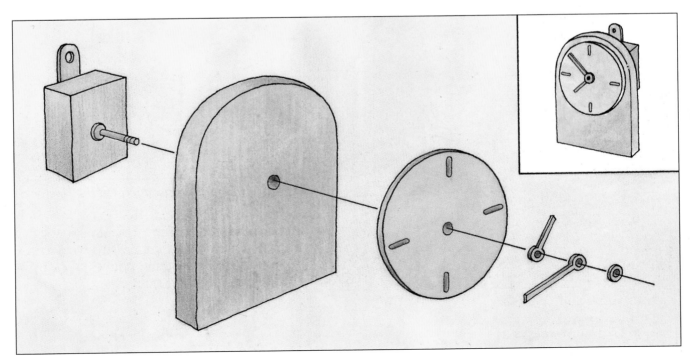

2 STRATEGIES

Modelling product performance in 3D

Three-dimensional models are a good way of exploring how a design idea will work as well as how it will look (see page 50). Sometimes these will be quick **sketch models** made with inexpensive materials. Sometimes they will be more complicated models made with construction kits.

Ergonomic modelling

In designing things for people to use we need to take the human form into account. The study of how easy it is for people to use objects and environments is called **ergonomics**. The ergonomics of your design proposal will show how safe, comfortable and easy to use it will be.

People come in different shapes and sizes and they move in different ways. If your model is life-size you can ask people to test it and record their responses.

Using easily worked materials like Plasticine or foam, you can adapt your model and find the design that suits most people.

Anthropometric data

Charts and tables can show average measurements for people in different age-groups for things like height, reach, grip and sight lines.

▲ *Card models used to test designs for handles.*

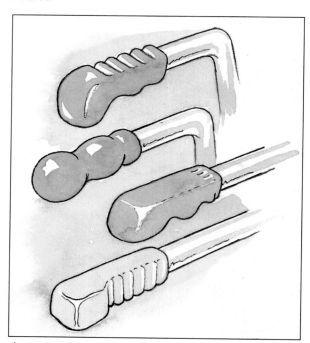

▲ *Plasticine models used to find the best grip.*

Such information is called **anthropometric data**. You can use it when you are making models to work out the details of, for instance, handles, or how your design might be held.

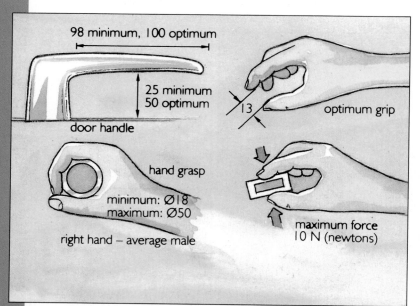

▲ *Making handles fit hands.*

Resource tasks
SRT 23–25

Modelling movement

Some designs involve moving parts and mechanical components like **cams**. Drawing these design ideas may not be enough. A working model can show actual movement and help you to see if your idea will work.

Remember:
- Keep your working model simple. Do not add unnecessary detail. Concentrate on the parts that move.
- Work on a scale that will let you model quickly. Tiny details of complicated parts are time-consuming to assemble. It may be easier to work on a much larger scale.
- The model should work in the same way as the final product. Choose materials and fixing techniques that behave like those you plan to use for the final design.

Materials

Card, paper and thin plastics are the most common materials used for this type of working model. They can behave like rigid materials, but are easy to cut and modify.

Pins, paper fasteners, and eyelet-type fastenings can be used as pivots. They allow movement and are easy to adjust.

You can also explore moving parts using easy-to-assemble construction kits (see page 61).

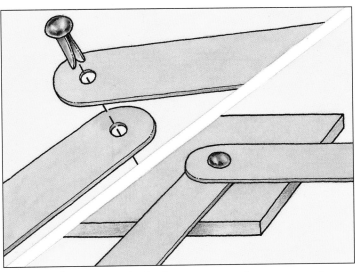
▲ *Making links between card components.*

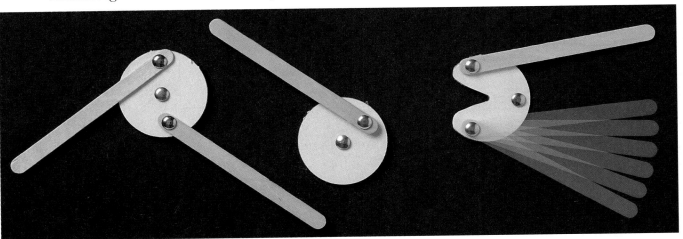
▲ *Models of different types of mechanisms.*

...Modelling product performance in 3D

Modelling structure

Many of the objects that you design will involve a structure. They may be based on a framework (like a pylon), a solid structure (a dam), or a shell structure (a domed roof). For all of these, **strength**, **stiffness** and **stability** are important.

A 3D model will show you very quickly whether your design is sound. By adding loads to your model, you can see how it resists stress, how stable it is, and you can discover the points of weakness.

It is also easier to identify the parts that are not functional, so they can be removed.

▲ *Two ways of making strong beams from card.*

Choosing the best materials

Structural models are usually small in scale, so it is important to use modelling materials that behave in a realistic way and show up problems in your design. Use a material that will show up lack of stiffness or weaknesses when tested. Materials that bend and snap easily will tell you about a structure under test situations.

- Suitable materials are paper, card, art straws, pipe-cleaners, spaghetti, cotton, balsa wood, paperclips and thin wire.
- For rapid joining use a hot-melt glue gun; PVA glue is stronger but takes longer to set.
- Only use adhesive tape in small quantities. If you use too much you may not be able to tell whether it is the structure or the tape that is strong!

Resource task
SRT 26

▲ *Modelling a crane structure with art straws.*

Modelling for fit

Sometimes you will need to find out how your design idea will fit *onto, into* or *around* something. For example, if you are designing something to be worn, a simple paper pattern is a good way of testing for size and shape.

Paper is easy to cut, fold and join, and can be modified time after time until the fit is correct.

A paper pattern is usually drawn to actual size so that tracings can be taken from it at a later stage in the design process.

Developments

Most packages or boxes are made up from flat sheets of card that have been folded up to make a 3D form. This is known as a **development** or **net**.

Open out a cardboard package to see how it has been folded. Notice the flaps that allow the box to be neatly joined together.

You can use developments for designing objects to be made from sheet materials. Paper or card developments allow you to see whether you have the correct size, shape and fit.

Simple forms can be fitted together to make more complicated ones.

If you are working on a scaled-down version of your idea, you can transfer the development to grid paper and enlarge the shapes to give you a pattern or template to draw around.

Resource tasks
SRT 27, 28

...Modelling product performance in 3D

Environmental modelling

When designing for an **environment** such as a bedroom, you will need to consider how objects are positioned within the space. It can be difficult to work out the impact of 3D objects on a space if you use only a 2D plan.

It is better to make simple, scaled, **block models** of the objects, and move them around on a plan to explore different layouts.

▲ *A set model used to plot movement.*

Block models can be quickly made from foam or card. You can use junk or found materials to give quick impressions.

Modelling like this will help you to imagine how people might use the space and move around within it.

▲ *A room layout using block models built to scale.*

Full-scale models

With small-scale models, it is difficult to get a true picture of how the layout will affect people. You can get a better impression with a full-scale model, using lightweight materials such as large empty cardboard boxes and fabric draped over poles.

You can test whether the layout works by getting people to move around in it and to pretend to perform tasks. Is there enough space for the doors to open properly? Can you walk comfortably between the units?

▲ *Full-scale modelling of an environment.*

Modelling with kits you can buy

Construction kits are designed to enable parts to be quickly assembled, taken apart and used again.

You can model structures, mechanisms, electronic circuits and pneumatic and hydraulic systems without having to spend time producing the individual parts.

As with any modelling, it is important to understand exactly what you are trying to work out before you start using kits like these.

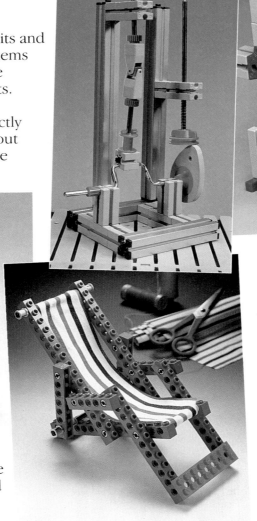

Mechanisms

Models using **gears**, **pulleys** and **axles** can be put together very quickly using a kit, so that you can experiment with different combinations of gear wheels or pulleys to achieve the results you want.

Electronic circuits

It is much easier to assemble circuits from ready-prepared 'systems boards' than with individual parts. The boards connect together simply and you can try out many different ideas quickly.

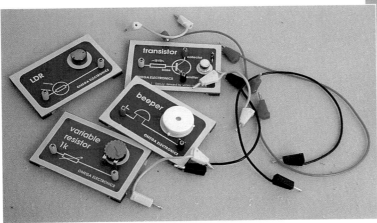

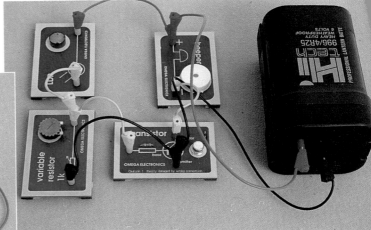

Resource tasks
MRT 12, ECRT 10–12

Modelling with computers

Modelling product appearance

Modelling on computers is sometimes called **computer assisted design (CAD)**. There are two main advantages to computer modelling:

- You can produce some sorts of drawings more quickly than with pencil and paper.
- You can make changes without doing the drawing again.

Thus you can build up a series of design drawings quite quickly and compare alternative designs more easily.

Most school software is *not* suitable for doing rough sketches and notes; it is much better to use pencil and paper for these.

However, it takes time to learn computer modelling and you will need to find time to practise.

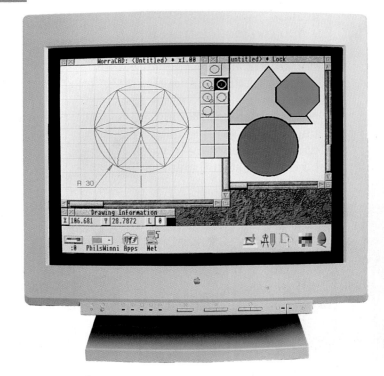

▲ *You can draw many shapes on a computer.*

Modelling shapes

You can use draw-and-paint programs to produce geometric shapes, curves or freehand lines.

Some software enables any part of the design to be enlarged or reduced in size, rotated, copied, or turned into a mirror image of itself.

Modelling surface decoration

You can easily use draw-and-paint programs to produce patterns.

You can select from a 'tool box' of pens, brushes or airbrush which can all be adjusted for thickness, spread and shape. Colour can be mixed on a 'palette' to fill in shapes or add shading or spray effects.

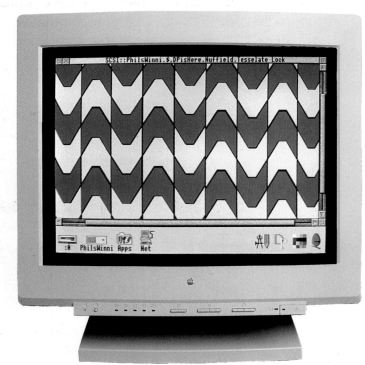

▲ *You can easily produce decorative patterns on-screen.*

Modelling form

You can use solid-modeller programs to produce 3D forms. These can appear on the screen as simple wire frames or solid objects.

Usually you can rotate these images to see them from all angles.

You can also colour the surfaces to show light and shaded areas to reflect the position of the light source.

▲ *You can experiment with different forms on-screen.*

▲ *Use a computer to get your work to look good.*

Modelling page layouts

You can use **desk-top publishing (DTP)** software to produce different layouts for a given text and illustrations. You can vary the number of columns of text, the size and shape of the illustrations, and the style and size of the lettering.

You can use a layout that you like as a template for every page in a document, making your work look professional.

Modelling built environments

You can use architectural modelling software to produce on-screen the image of an interior design. You can experiment with this design until you find the most suitable model. You can alter the proportions of a room, making it taller or wider. You can alter the colour scheme. You can experiment with different furniture layouts.

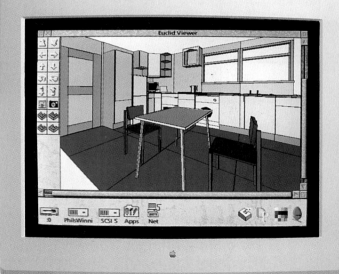

▲ *You can explore different interior designs with a computer.*

...Modelling with computers

Modelling product performance

Modelling how your design will work using computers (CAD) will save time.

Modelling mechanical action

You can design mechanisms on-screen and see whether they work as they need to for your design.

You could do this modelling using a construction kit, but it would probably take longer and you would be limited to the parts in the kit. On-screen a wider range of parts is available and you can even design your own.

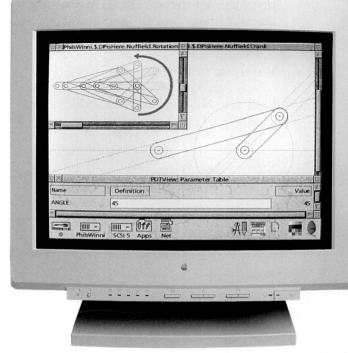

▲ *Modelling the action of a crank, link and slider on-screen.*

Modelling the behaviour of electronic circuits

You can design **circuits** on-screen and see whether they do what you want. Once you have designed a circuit that works on-screen you can build it knowing that it will do the job. Using an electronics kit would probably take longer and you would be limited to the parts in the kit.

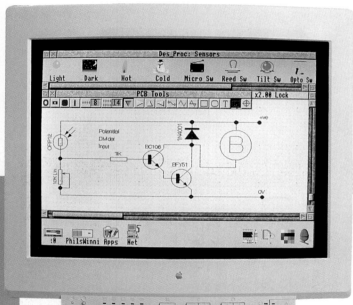

▲ *Testing circuits on-screen.*

Modelling energy transfers

The need to save energy is particularly important in the design of buildings, where heating uses a lot of energy.

It is quite easy to find out just how much energy would be saved by using double-glazing, draught-proofing, roof insulation, different construction materials and even changing the shape of a house by using a computer.

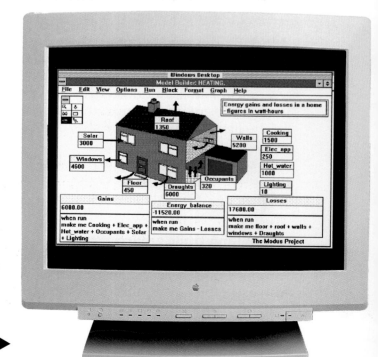

Testing energy efficiency on-screen. ▶

Modelling nutrition

You can use a nutrition database to identify and calculate the food values of ingredients for meals. You can compare these values with the dietary requirements of people eating the meals.

Modelling ergonomic performance

It is easy to show different workspace layouts on-screen and to draw in the movements needed to carry out particular jobs. In this way, the most efficient layout, the one that requires least movement, can be identified.

To do this with pencil and paper would take much longer.

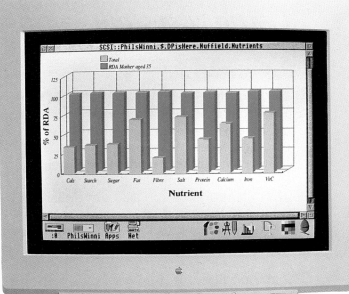

▲ *Using a computer to check on a diet.*

▲ *Finding the best layout using a computer.*

Modelling business performance

When you are developing a business plan, it is important to be able to model how money moves in and out of the business.

You can use a spreadsheet to help you do this. Put in figures for all your costs and expected sales, and in just a few seconds the spreadsheet will be able to calculate whether you make a profit or loss. Doing all the necessary calculations by hand would take a very long time.

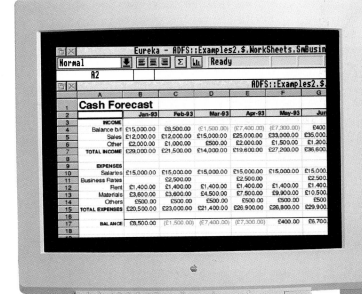

▲ *Using a computer to check on money movement.*

Using a systems approach

A way of thinking

You can think about anything that is made up of a collection of parts that work together to do a job as a **system**. The parts of the system might be objects, people or both.

Systems thinking works best if you are trying to understand something complicated. It is best to think about a torch as a simple circuit made up of a battery, bulb and switch. But it will help you to understand a radio if you think about it as a system made up of different parts – aerial, tuner, amplifier, loudspeaker, etc. – that work together. Systems thinking helps with the design of complicated products.

▲ *It is best to think about this radio as a system.*

The system boundary

The **system boundary** is like the borders of a country. Everything inside the boundary belongs to the system. Anything outside does not. However, unlike the borders of a country, with a system boundary you can choose to put the boundary where it helps your understanding or design thinking.

If you were designing just the internal workings of a radio your system boundary would not include the case, battery-changing arrangements, tuning dial or switches. If you were designing the case as well you would draw the boundary to include all of these. So whenever you are designing a system the first thing you need to do is decide on its boundary.

Some unusual system boundaries

Gas fire and boiler manufacturers must include the house as part of their system.

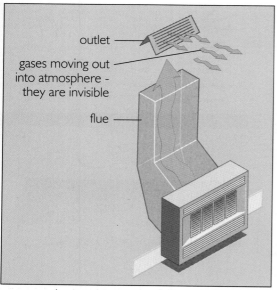

▲ *Waste and poisonous gases must be removed from the house.*

▲ *Cinema owners include nearby houses in their system boundary.*

Systems and subsystems

To understand a system you can divide it up into simpler **subsystems**. You then work out how these need to be connected together.

A music centre can be divided into several subsystems – such as the record deck and the tuner. In some music centres these subsystems might be in separate units. In others, they might all be in the same box.

You can draw a **system diagram** showing how the subsystems are connected together to make the complete system. Information in the form of electrical signals flows along the lines connecting the different subsystems.

Note that this system diagram is not meant to look like a music centre. But it does help you to understand how a music centre works.

▲ *A music centre and its system diagram.* ▼

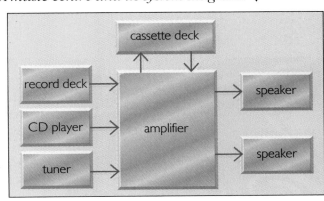

Inputs and outputs

A designer beginning to design a complex product using systems thinking only needs to consider what goes into the system and what comes out. She does not need to worry about what happens inside the system just yet. All the things that go into the system are called **inputs** and all those that come out are called **outputs**.

▲ *Some of the inputs and outputs of a fast-food service.*

The designer of a fast-food service might identify the inputs and outputs shown here. Next he will break down the system into subsystems and see how the inputs and outputs need to be arranged. A system diagram showing these inputs and outputs and some of the subsystems is shown on the right.

You should be able to see that the output of one subsystem becomes an input of another.

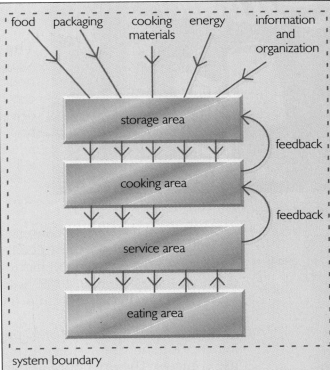

▲ *A fast-food service as a system diagram.*

Resource task
SRT 29

...Using a systems approach

Connecting systems to the outside world

The user

Many systems need to be used by people who have had no special training. For example, shoppers need to be able to tell the system what they want, collect their orders and pay. Or someone using a bank needs to be able to ask for and get money and information. Users do not need to know how a system works, just how to use it.

The parts of the system that they see, touch, talk to or handle are called the **user interface**. Sometimes this is another person, like a shop assistant or bank clerk. The advantage is that it is friendly. Users can easily ask questions if they are not sure what to do.

◀ *Machine and human* ▶
interfaces.

Human interfaces need special training to enable them to deal with user enquiries. The disadvantage is that they are not available 24 hours a day. Often the user interface is a machine, as in a soft drinks machine. There are written instructions to read and buttons to press instead of someone to talk to. The advantage is that it is always available. The disadvantages are that it can appear unfriendly and difficult to use. The user interfaces for such machines must be designed so that they are self-explanatory and easy to use.

The operator

Many systems need to be controlled by an operator. The operator of a system needs information about the system and a way of controlling it. For example, a car is controlled by its driver. The car driver needs to know how fast the car is going, how much petrol it has, and other detailed information. She needs to be able to increase the car speed, slow down and turn left and right.

The parts of the system that the operator looks at, touches, talks to or handles are called the **operator interface**.

These are usually more complicated than user interfaces. An operator needs more information than a user and has to be able to do more things. Operators are usually trained to operate the system, while users are not.

The operator must be able to put information into the system through easy-to-use controls. The operator interface should be as easy to use as possible.

The operator interface of a modern car. ▶

Resource task
SRT 30

Feedback

Most systems need to respond to changes. A central heating system needs to be able to turn itself off if a room becomes too hot and on again when it becomes too cold.

To do this the system needs information. This information is the temperature in the room. It could be provided by a temperature sensor.

This type of information is called feedback. Systems with feedback are called **closed-loop systems**.

A system without feedback is called an **open-loop system**. A room heated by an electric fire is an open-loop system. If the fire is left on it will carry on heating the room even if it is a hot day. There is no feedback telling the fire that the room is too hot and to turn itself off.

System diagrams for both of these systems are shown below.

▲ Closed-loop and open-loop systems. ▼

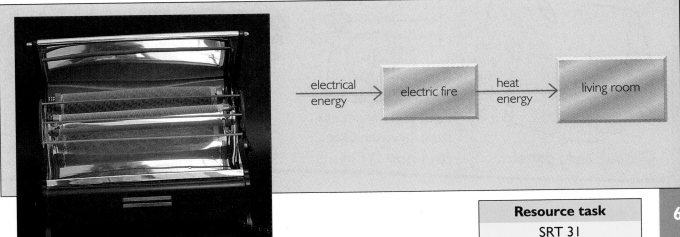

Resource task
SRT 31

Planning techniques

Sequence diagrams

Sequence diagrams show the order in which events happen.

Flow charts

A flow chart is a sequence diagram using single words or short statements to organize a task or event into a series of separate steps.

Storyboards

If you want to make a sequence diagram using pictures – a storyboard – you may need to start with a flow chart in order to answer the following questions:

- How much information is needed?
- How many stages are involved?
- What is the correct order of the stages?

This type of diagram is to help you to work out a sequence of events and is not designed to be an illustrated guide for a wider audience (see User support, page 96).

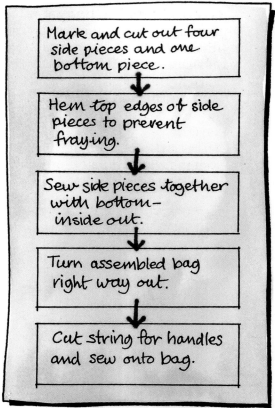

This flow chart describes a student's plan for making a carrying case for school equipment. ▶

▲ *A storyboard using simple line drawings to show the stages in producing a glove puppet.*

Gantt charts

Gantt charts are the most useful way of planning when several different things are going on at once.

The Gantt chart below shows the planning for a puppet play performance. Several pupils are involved. They know that they have to write the script, design and make the puppets, the theatre and scenery, develop any special effects, and rehearse the play for a performance in six weeks' time. Have they allowed enough time for all they have to do?

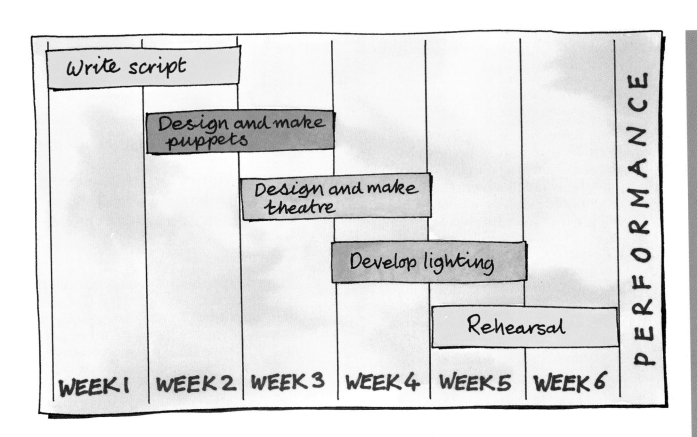

Evaluating outcomes

The user trip

▲ *Whatever the product you can take a user trip to evaluate it.*

Four different outcomes of design and technology are shown on this page. The supermarket is a complex system. The telephone box is a complicated communications device. The tin opener is a simple tool, and the chicken tikka masala is a convenience food.

When you try to evaluate an outcome you are trying to find out its good and bad points. A simple way of doing this is to take a **user trip**. This just means using the product in an ordinary way and asking yourself the following questions:

- Is it easy or convenient to use?
- What is its job and does it do it?
- Do I like it?
- Would I want to own it or use it?

You would evaluate the four outcomes like this:

- the supermarket by going shopping there;
- the phone box by making a telephone call;
- the chicken tikka masala by heating and eating it;
- the tin opener by using it to open a tin.

While on the user trip you note down the answers to the questions. You can use them to make a list of 'improvement suggestions'.

Sometimes it helps to get other people to take user trips and for you to observe them and then ask questions. In this way you can collect views from users with different viewpoints. For example, a tin opener that works well for most of us may be quite difficult for someone with arthritis to use.

Resource task
SRT 32

Performance testing

To evaluate an outcome you need to find out whether the product does what it is supposed to do. The specification will tell you what it has to do. So you need to compare what it does – its performance – with the specification. Here are two examples.

A child's hat

Look at the specification for a hat for children 3–5 years old. The designer tried to meet these requirements by providing an adjustable head strap plus a chin strap with a Velcro fastening. He used PVC for the outer covering.

To test the hat's performance you need to:

- try it on a range of different-sized heads to see whether it fits them all;
- observe children wearing it on a windy day to see whether it comes off;
- spray water on it to see whether it comes through to the lining;
- watch children trying on the hat to see if they can manage.

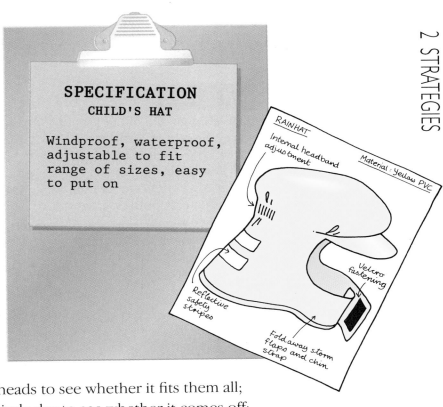

SPECIFICATION
CHILD'S HAT

Windproof, waterproof, adjustable to fit range of sizes, easy to put on

A toy car

Look at the specification for a toy car for children 5–8 years old.

The designer tried to meet all these requirements by designing a hand-held control unit which is attached to the toy by a ribbon cable. The control panel has an on/off switch to control the lights and switches to control the motor. The large tyres have a deep tread to provide the grip for climbing. The battery is housed in the control unit so that the motor only has to move the toy and not the battery as well.

A performance test will need to answer these questions:

- Can the car go forwards and backwards?
- Will it climb over obstacles?
- Will the headlights turn on and off?
- Can children use the hand-held unit to control the car?

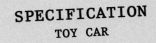

SPECIFICATION
TOY CAR

Travel forwards and backwards, climb over obstacles, small electric motor, headlights turn on and off, suitable for 5-8 year old

2 STRATEGIES

...Evaluating outcomes

Winners and losers

Any technological change will affect lots of people, some directly, others indirectly. Some will gain from the change and others will lose. As part of your design and technology work you will need to evaluate your ideas and products. Identifying winners and losers will help you decide how good the designs are.

A pupil evaluated his design for a disposable waterproof hat for hikers, using a **target chart**.

1 He listed all the people who would be directly affected if his design were to be mass-produced. They included:
- hikers;
- those who own or work in the factories making the hats;
- those producing the plastic material;
- the shop keepers who would sell the hats.

These were written on the first ring of the target.

2 He thought of others who might be affected indirectly, such as:
- those manufacturing rival products;
- people who might find the hats thrown away on their land;
- those affected by the plastics industry, and so on.

He wrote these on the next ring.

3 He looked at each group of people and decided whether they were winners or losers. He highlighted the winners in yellow and the losers in blue. By looking at the balance of blue and yellow colours in each ring he was able to use his target chart to help him evaluate his idea.

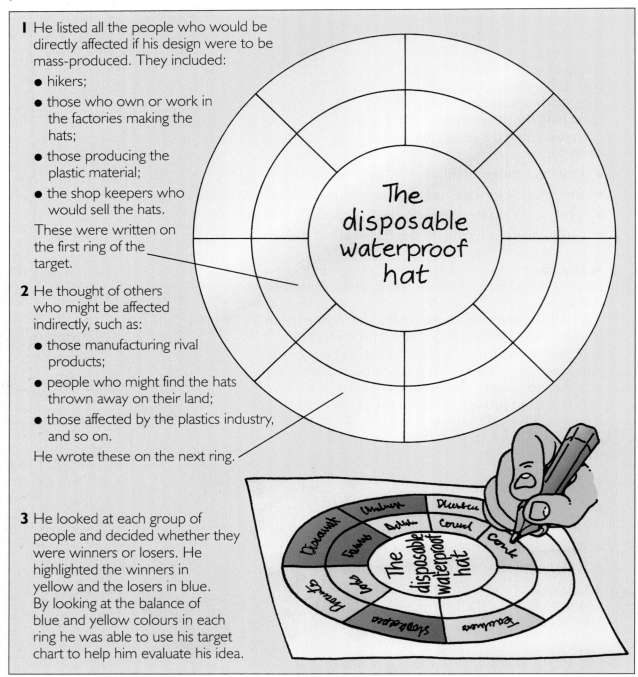

Resource task
SRT 33

Is it appropriate?

Appropriate means 'suitable', so appropriate technology is suitable technology.

Is it appropriate?

Technology is appropriate if …	You can check whether appropriate technology is being used by asking these questions:
… it suits the needs of the people.	Is it what the people need and want?
… it uses local materials.	Does raw material need to be transported?
… it uses local means of production.	Do local people make it near where they live?
… it is not too expensive.	Can the people afford to buy, run and maintain it?
… it generates income.	Are jobs created or people made redundant?
… it increases self-reliance.	Does it improve people's lives?
… it uses renewable sources of energy.	What fuels does it use?
… it is culturally acceptable.	Does it fit in with the way the people live?
… it is environmentally friendly.	Does it damage or improve the environment?
… it is controlled by the users.	Does it need outside experts?

Very few types of technology will score highly against all these questions.

Whether these tools are appropriate forms of technology will depend on the situation. ▼

Resource task

SRT 34

3 Communicating design ideas

Why and how?

This chapter looks at ways of communicating your design ideas to different audiences.

Before starting any drawing or model during a design and make task you need to be clear about:

- the purpose of the drawing or model – what it has to communicate;
- the audience – who you are trying to communicate with.

Visualizing and developing ideas

In the earliest stages of designing, **thumbnail sketches** help you to visualize and record ideas for yourself.

Later, you will need to communicate your ideas to others. For this, the drawings are likely to be **rough sketches** that you need to explain for your ideas to make sense. In a design team, this may be an important stage in the development of your design.

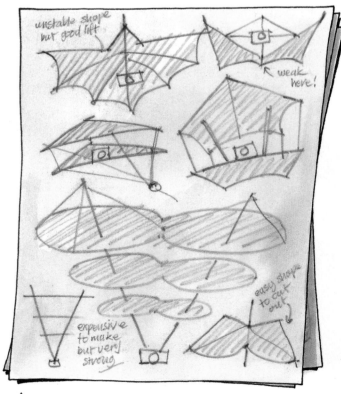

▲ *Thumbnail sketches help you to visualize your design.*

'Selling' your idea

When you are happy with your design idea you need to 'sell' it to persuade the client (or teacher) that it is worth producing. For this you make **presentation drawings** or **models**. These should show how your idea will look and explain how it will work.

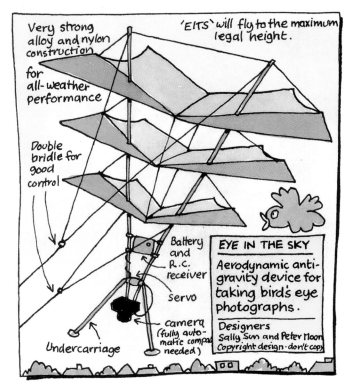

▲ *A presentation drawing.*

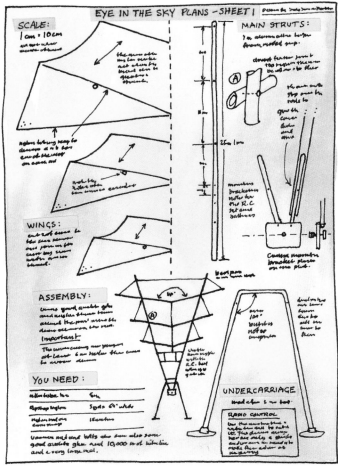

▲ *Working drawings.*

Drawings for making

Before a design can be made, a 'drawing for making', a **working drawing**, is needed. It must communicate the exact details of the design.

It specifies the materials needed, the shapes and sizes of all the parts to be made and how these fit together. Someone should be able to make the design from your working drawing without having to ask you any questions.

Informing the user

Often, you will need to give information to the user of your design. You might need to make a model showing how the design works, to produce instructions explaining how to use it or a guide to repair and maintain it.

You may be involved in marketing, producing materials to persuade people to buy your product.

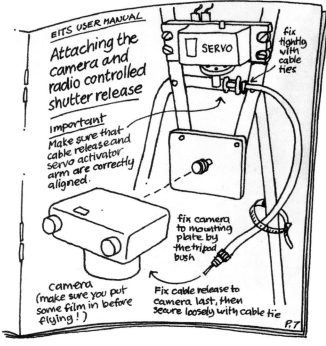

▲ *Diagrams can help others to use your design.*

Presenting your product idea

You can produce presentation drawings in many different ways. Three examples are shown here.

It may take several drawings to present your idea well. Just one of these bicycle alarm drawings would not be as effective a presentation as all three.

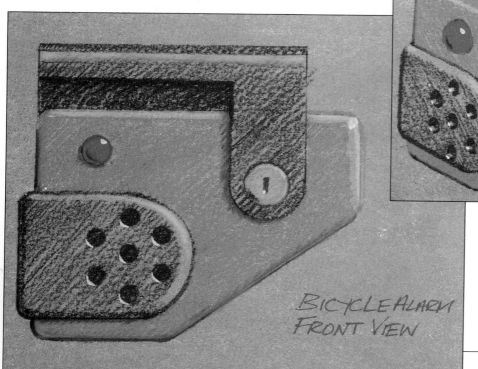

BICYCLE ALARM FRONT VIEW

BICYCLE ALARM

BICYCLE ALARM SEE-THROUGH VIEW

OOEEOOEEOOE

Using isometric views

An **isometric view** is a way of showing three dimensions on a drawing. You can use special grid paper.

- Draw the object at an angle, with one corner as the closest point to you.
- Draw all vertical (upright) lines on the object as vertical lines on the drawing.
- Draw all lines which are horizontal on the object at 30° to the horizontal on the paper.

These step-by-step drawings show how the isometric view of the camera is constructed. You can use the crating method (page 48) to add details. Practise drawing isometric views on grid paper before you try to draw them freehand.

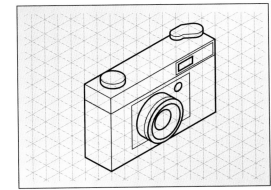

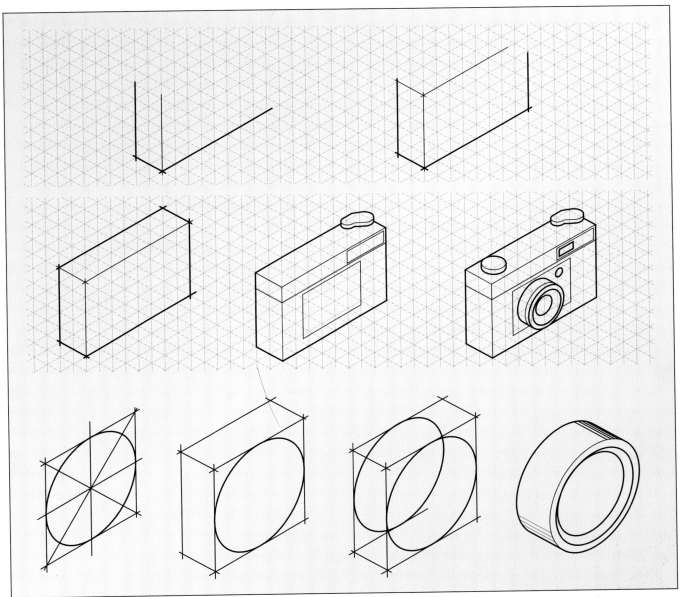

Resource task

CRT 1

...Presenting your product idea

Giving surfaces texture

Choose the best view of your product. Then think about how you can colour the drawing to imitate the material the product will be made from, and the finish you intend to give it.

Plastics, glass and metals often have shiny surfaces. We see these surface finishes in the way they reflect light. Show this by adding highlights and reflections to your drawing using the techniques shown here.

Materials such as rubber, card, wood and some plastics often have only a dull surface and do not show harsh reflections or highlights. Show this on your drawing by applying an even tone of pencil, pastel or marker, and adding soft-edged reflections and highlights.

The texture is not applied evenly, but more heavily on areas that are in shade, and sparingly on areas facing the light source.

Another way to convey surface texture is to place your paper over a real textured surface and rub lightly with a pencil or crayon. Try this with glass paper, hardboard and fine wire mesh.

Resource task
CRT 2

Giving surfaces depth

Step 1

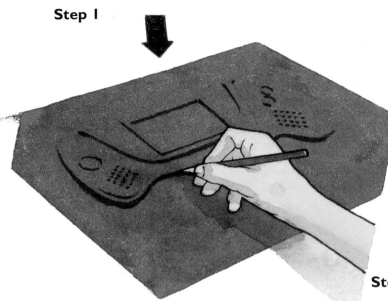

You can convey depth and 3D on a flat view of an object by adding highlights and shadows. You need to understand how light falls onto an object and creates areas of light and shade.

This flat view of a games console has added highlights and shadows to give depth.

This technique works best on coloured paper, as you do not have to colour in the whole surface, just the shadows and highlights. If you use white paper you will need to colour the body of the object with markers or water colour.

Step 2

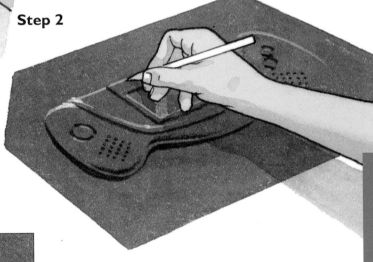

Follow these steps:

1 Decide where the light is coming from. Darken all the edges facing away from the light.
2 Use a white pencil-crayon to lighten the edges facing the light source.
3 Add sharper highlights using the crayon or a fine paint brush and white poster paint or gouache.

Step 3

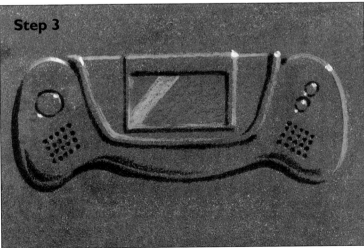

4 You may need to include more views to show the shape of the product.

Step 4

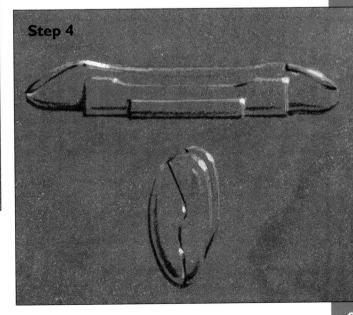

Resource task

CRT 3

3 COMMUNICATING DESIGN IDEAS

Presentation drawings for textile designs

Your design work with textiles will often involve clothing and fashion accessories. These are nearly always presented as if in use, to convey size and proportion, and show how the design is meant to be worn or carried. It can also suggest the image and effect that the design creates as a fashion item.

Drawing people can be difficult. There should be spaces between the clothing and the body – unless you are designing a skin-tight item like a swim suit! Here are some hints.

Trace over a photograph of a model in a magazine. Draw your design over it.

Simplify heads, hands, feet and the body, if you are designing items for them.

Fashion drawings sometimes show models with false proportions. The head may be drawn small, the shoulders wide, and the arms and legs extra long. This gives an elegant, stylish effect.

An outline of a figure will be enough if you only want to present an idea for an item like a rucksack.

Resource task
CRT 4

Using different media

Your presentation drawing should show the most important features of your design. If you have given special attention to details, include a magnified close-up (see page 54).

Attach samples of materials, called **swatches**, buttons and small samples of decoration to the drawing to give a more detailed picture of the final design.

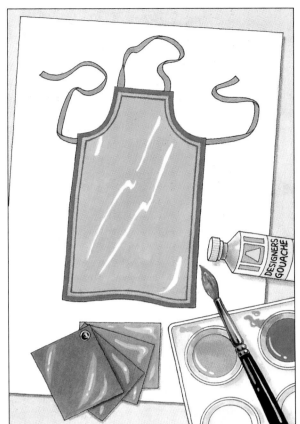

Your presentation also needs to suggest the 'feel' of the material used in your design. Mix different media to obtain the effect you want. Try these:

- **water-colour paints** – use water-resistant wax crayon before using the paints to make a pattern;
- **gouache** – good for strong, flat colours;
- **pencil-crayons** – good for hinting at detail and texture on top of colour washes;
- **pastels** – give a soft, matt effect, especially if rubbed on with a tissue or cotton wool;
- **brush pens** – give strong, flowing outlines to clothes;
- **felt-tip pens** – for fabric designs and decoration if you can get a good range of colours.

Resource tasks
CRT 5, 6

Presenting food product ideas

The best way of presenting a food product idea is to prepare the real thing. Drawings and models lack taste and smell, which are so important.

In industry, home economists usually present a number of variations on a product for tests involving taste, appearance and texture. The test results could form the basis of a presentation along with photographs of the product and descriptions of the content.

A video record of taste tests or a survey could also be used.

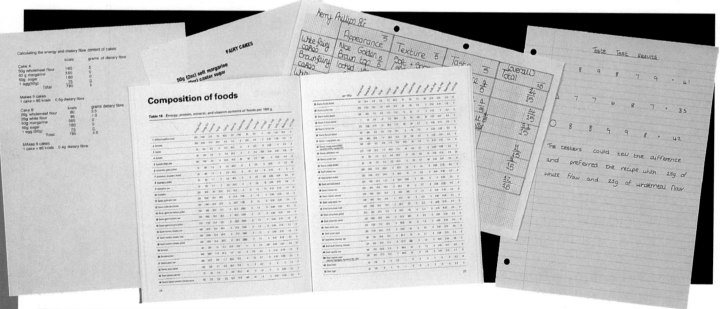

Packaging and advertising are important, as are data connected with nutrition and user preferences (see page 224).

Photographs on the packet, and in cookery books and advertisements, present food in the most attractive way possible.

You may need to use photographs or a videotape recording to keep a record of the appearance of the food products that you design.

Presenting your idea in a context

Backgrounds

Your presentation drawing may have much more impact if the product is shown against a background, rather than 'floating' on a sheet of white paper. Even a simple frame attracts the eye. Here are some ideas.

Draw a frame around or under the image.

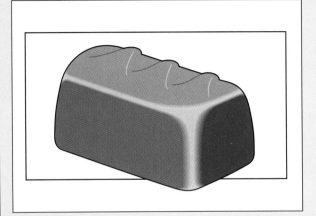

Add a shadow under the object. Use pencil or a grey marker.

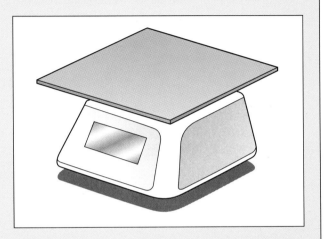

Draw a rectangle behind the image. This may suggest a horizon line or a landscape.
   Cut out the shape you want from coloured paper or use pastels with masking tape for a clean outline.
   Paste a cut-out of your design drawing over the top.

Place it into a 'real' environment. You could use a ready-made magazine picture or one of your own photographs.
   Paste a cut-out of your drawing on the picture, or on a tracing paper overlay if the drawing is 'lost' against the background.

Resource task

CRT 7

...Presenting your idea in a context

Putting it in a picture

For your presentation, you may want to show your design in a drawing of the place where it will actually be used. A technique often used to do this is **perspective drawing**. It gives an impression of depth as well as height and length, as we see them in real life.

In the picture the cars and lamp posts seem to get smaller, and the road to get narrower until it disappears at a point. This is known as the **vanishing point**.

A perspective drawing recreates this illusion on paper. **Single-point perspective drawings** are often used to show interior designs, landscapes and stage sets.

To draw a one-point perspective view of a room, follow these steps:

1 Draw a line to represent the horizon. This line is known as the **eye level**.

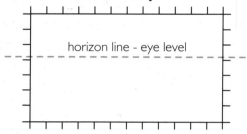

2 Mark a vanishing point on the horizon and draw in lines to it from the corners.

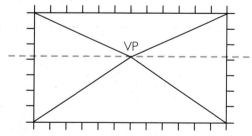

3 Estimate the depth of the room and draw in the back wall.

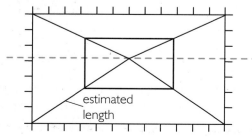

4 Draw in guidelines using a scale taken from the front of the frame.

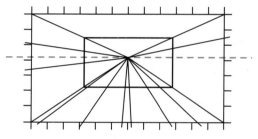

5 Add details and rub out the guidelines.

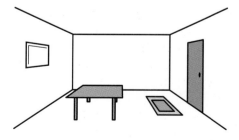

▲ *A finished single-point perspective view.*

Resource task
CRT 8

Displays and exhibitions

An important part of your presentation may be a display.

Whatever the product, the most important question you can ask yourself is, *'How can I get my ideas across?'*

Presentation models

A good way to put over your idea may be to build a presentation model or **prototype**. These are accurate, detailed models that show exactly what the final design will look like, and sometimes how it will work.

They can be made of any material that is easy to work with and will give a realistic finish. A few coats of spray paint and the addition of details like switches and lettering can help.

◀ *This 'block model' of an iron is made of polystyrene foam but gives a good idea of how the finished product would look.*

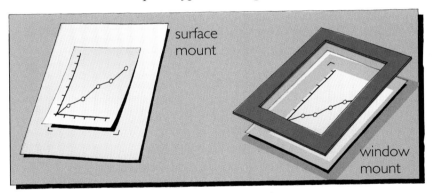

▲ *A prototype working model of a toy car.*

Mounting a display

If your presentation consists mainly of 2D work, consider mounting it for display.

An easy way is to **surface mount** the work, fixing it on to sheets of mounting paper or card.

Or you can **window mount** the work behind a mounting sheet with a hole or 'window' cut in it.

Choose the colour of the mounting material to show your work to the best effect.

Add headings, labels or captions to help explain the display. Lettering should be neat and easily understood (see page 88).

When you mount sheets together, keep the layout simple. One technique is to line up one or two edges of a sheet with others on the display. ▶

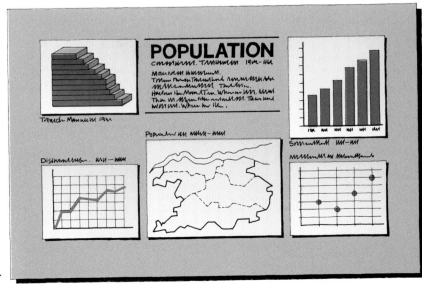

3 COMMUNICATING DESIGN IDEAS

Presentation reports and documents

If written material forms an important part of your presentation, it should be laid out clearly and attractively to help others understand your proposals.

Lettering

Using computers
Choose lettering styles that are easy to read. These can often communicate as much as the words themselves. One style may suggest seriousness while another may give a more light-hearted impression.

There are many different designs of lettering – called **fonts** or **typefaces**. If you use more than two or three on one document it can look messy.

Hand lettering
Poor lettering can spoil your design presentation. Any hand-written notes, titles, headings or labels should be clear and easy to read.

Practise hand lettering by drawing guidelines. These help keep your writing in a straight line and form each letter to the right proportions.

You can also use dry transfer letters or a stencil. These are good for small lettering tasks but take too long for general work.

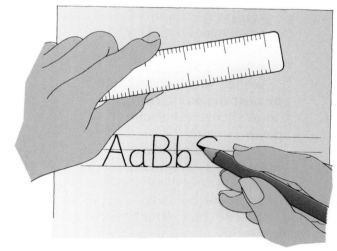

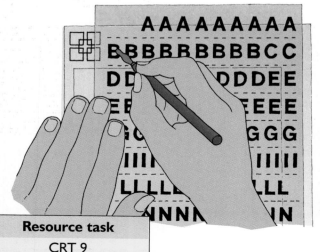

Resource task
CRT 9

3 COMMUNICATING DESIGN IDEAS

Photographs

Photographs can record the progress of a design task. They may help to explain how you arrived at your proposal.

With an overlay of tracing paper or clear film, a photograph can be used to show proposed changes. ▶

▲ *Photographs can be pasted together to show a panoramic or composite view.*

Layout

Layout is the arrangement of words and pictures on a page.

Designers may use a **grid**, which lets them place words and pictures together in different ways, but still keep the same basic 'look' for each page.

If you are using a DTP or word-processing system, investigate the use of grids to present your work. Some examples are shown here.

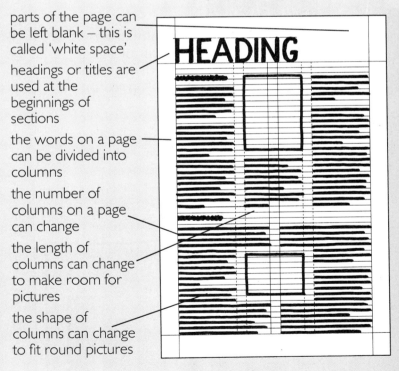

- parts of the page can be left blank – this is called 'white space'
- headings or titles are used at the beginnings of sections
- the words on a page can be divided into columns
- the number of columns on a page can change
- the length of columns can change to make room for pictures
- the shape of columns can change to fit round pictures

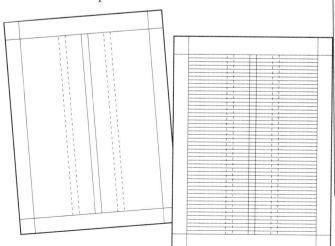

Resource task
CRT 10

Using facts and figures in presentations

Facts and figures collected during your design research should be presented accurately but in an attractive and clear way.

You can do this by presenting them as pictures or **graphics**. Here are some ideas.

Graphs

A graph can show how something changes over a period of time. For example, you could use one to show temperature changes during a day.

If you have more than one line to plot, code each in a way that clearly identifies them.

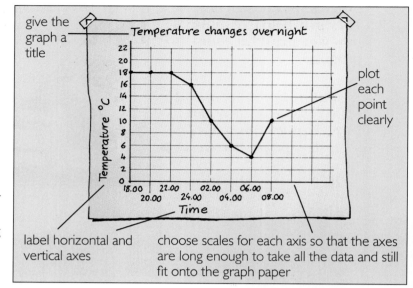

This gives the same information as the other graph but in a more interesting and attractive form. ▶

Bar charts

A bar chart is a way of comparing different amounts. The simplest bar chart is a single column divided up to show each item as a proportion of the whole.

Different amounts can also be shown as vertical or horizontal blocks drawn to scale along an axis.

Whichever method you use, choose a scale that makes it easy to 'read off' the amount accurately.

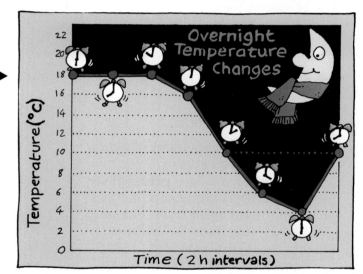

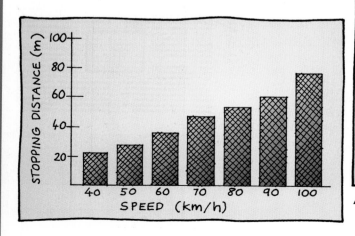

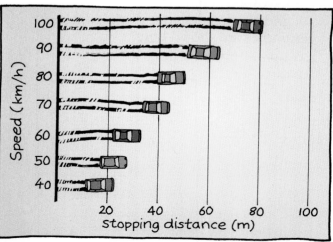

▲ *This bar chart has more visual impact than the one on the left.*

Pie charts

A **pie chart** is a good way to present data graphically, when you want to show the proportions of the parts making up a whole.

It is usually a circle divided into portions like the slices in a pie. The circle represents 100% – the whole. The size of each portion shows its percentage of the whole.

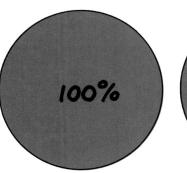
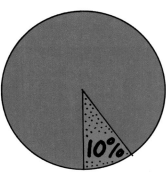

To draw a pie chart:

- work out the percentage size of each quantity;
- turn this percentage into an angle: 100% is 360°, so 10% would need a 36° slice, 15% a 54° slice, 25% a 90° slice and so on;
- label each slice and add the percentages or values so that it is clear what each represents.

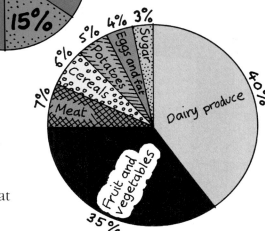

▲ *A pie chart comparing food purchases.*

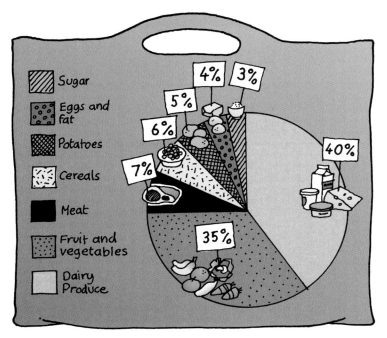

▲ *A pie chart must present information accurately and clearly, but it will have more impact if it is drawn in an interesting way.*

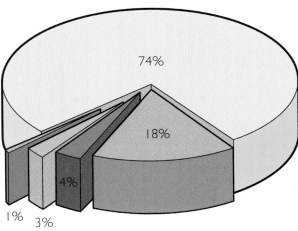

▲ *You can use computer programs to do the calculations and draw pie charts for you. Some programs will draw the pie chart in 3D.*

Resource task

CRT 11

Information for making

When your design proposal is ready to be made, you need to put together the 'information for making'. This includes all the working drawings and instructions that you, or someone else, need to make the design.

Putting this information down on paper is also a useful check on how well you have thought through your idea.

Recipes

You will need planning tools when you are working with food (see page 70), but also you will probably want to record your design idea as a detailed recipe.
The recipe lists the ingredients, and gives step-by-step instructions on how to make the product.

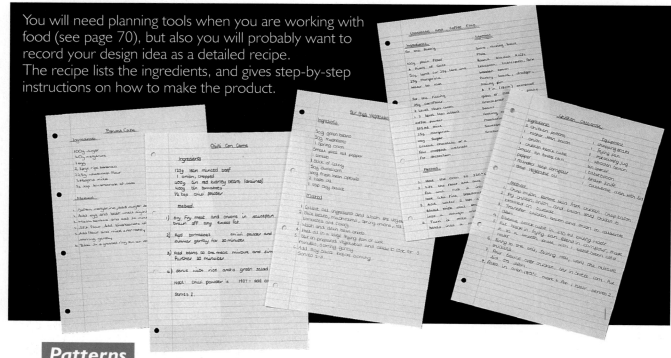

Patterns

When working with textiles, it is helpful to draw patterns on squared paper, especially if they need to be enlarged or reduced (see pages 138–9).

Mark allowances for joining materials on your patterns.

Special fixings or fastenings need to be highlighted, using sequence diagrams if necessary.

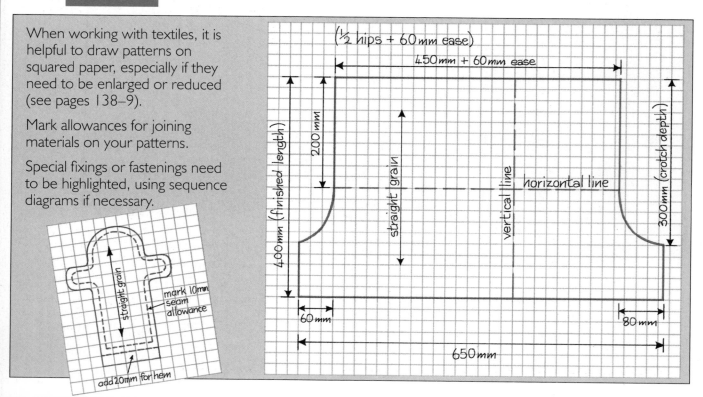

Orthographic views

Working drawings usually show some square-on views (or elevations) of the product that is to be made. These views, drawn as a related group, are known as an **orthographic projection**.

The diagram shows the six views obtained by looking at the camera in the direction of each arrow.

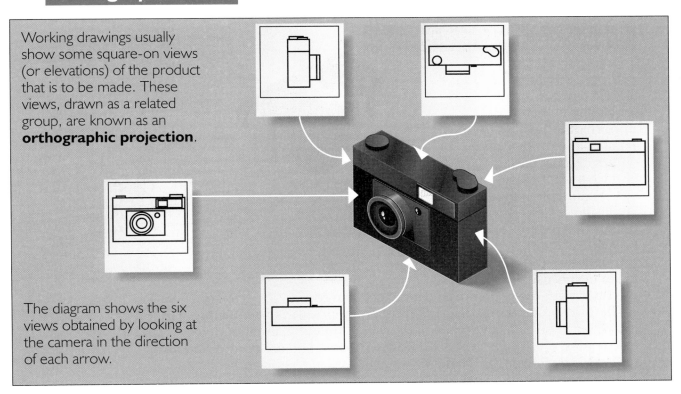

You need only draw three views to give enough detail about your design. The names given to each are:
- **the plan** – the view looking down;
- **the front elevation** – the main view giving the most information;
- **the end elevation** – the remaining side view(s).

Each view must be drawn so that its relation to the others is clear. Using grid or graph paper makes it easier to line the views up with each other.

It can also help to imagine that the product is suspended inside a transparent box. Draw the views as if traced on the sides and top of the box, which is then opened out flat.

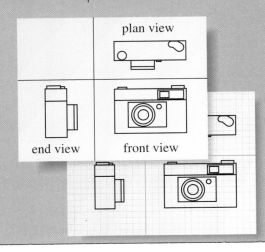

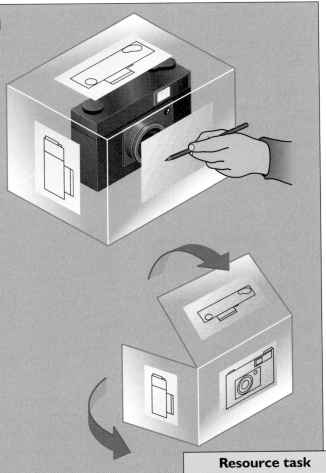

Resource task

CRT 12

... Information for making

Producing working drawings

To make an accurate orthographic drawing, you should use a drawing board and instruments. It is important to be able to draw parallel lines and accurate angles, as part of a neat, clean and clear drawing.

A working drawing must show all the dimensions and details needed for a product to be made.

In industry the design and make task is usually shared, so it is important that a common 'language' is used.

The British Standards Institution recommends ways of showing information on different types of drawing – building, engineering, electronics, etc. These are called **conventions**, and are recognized and understood all over the world.

A complete working drawing includes:
- exact sizes with measurements, usually in millimetres, for each part;
- the materials to be used for each part;
- detailed construction and assembly instructions;
- surface finish required for each part;
- a complete parts list.

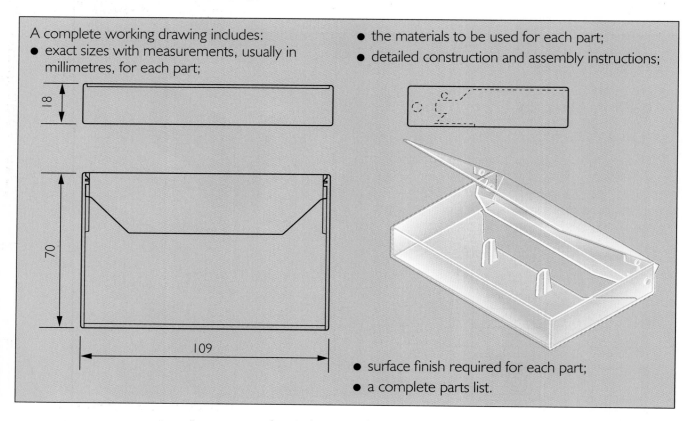

It may not always be possible to make a full-size drawing of your design. It may be too large to fit on the paper, or so small that you would not be able to see the detail.

Choose a suitable scale and write it on the drawing. A full size drawing is 1:1, an object drawn at half size is 1:2, and one shown at twice the size is 2:1.

Resource tasks
CRT 12, 13

Assembly details

Most products you design are made up of several **components** (parts). On a working drawing each must be identified and shown in its correct position for **assembly**.

A good way of doing this is by drawing an accurate 'exploded' view of the product (see page 55). These views may be 2D or 3D. Label or number each component to help with identification.

PARTS LIST
1 Shell top × 1
2 Auto-shutter × 1
3 Write protect slider × 1
4 Liner × 2
5 Metal hub × 1
6 Disc × 1
7 Shell bottom × 1

Parts list

A parts list is a table of information. For each component it gives:

- part number;
- description;
- quantity needed.

If it is on a working drawing it will also list:

- size;
- material it is made from;
- surface finish.

Resource task
CRT 14

3 COMMUNICATING DESIGN IDEAS

3 COMMUNICATING DESIGN IDEAS

User support

Now you have designed and made your product you need to persuade someone to buy it.

You will need to make sure that the person using it understands how to use and look after it. For example, a new oven will come with instructions on installation, a description of the controls and their use, and guidance on cleaning and safe use. There may be a booklet giving guidance on cooking different foods and some recipes.

The user might have bought this cooker because a model showed how new technology had been used to improve its efficiency. Or information in an advertisement or consumer report may have helped him or her to choose it.

Different sorts of user support. ▶

Look at the following pages to decide what kind of user support might be right for your product.

Demonstration models

Demonstration models are used to show exactly how a final design will work. They can help to explain an idea, a process or a system, often using a simplified version of the actual design.

Sometimes it is helpful to scale-up the model (for instance, a small mechanism). At other times the model can be scaled-down (for instance, a hydro-electric plant).

The model may be a graphic representation of a system or process, and may take the form of a diagram, such as a flow chart (see page 70).

Small-scale model of a wind turbine on a farm. ▶

Guides and instructions

Many products are supplied with a **user guide**, a set of instructions to ensure that the product is used effectively and safely. This information must be clear and easy to understand, so care needs to be taken in the production of the guide.

Communicating information graphically – using pictures or diagrams – can be very effective.

Step-by-step instructions for some self-assembly furniture. Note the use of 'exploded' views and no words. ▶

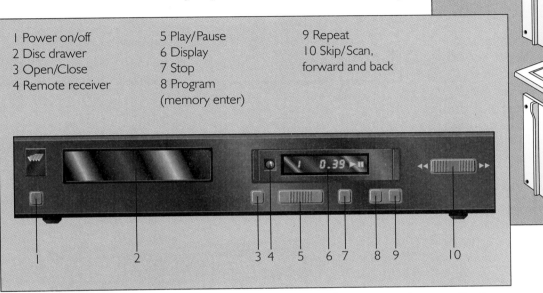

1 Power on/off
2 Disc drawer
3 Open/Close
4 Remote receiver
5 Play/Pause
6 Display
7 Stop
8 Program (memory enter)
9 Repeat
10 Skip/Scan, forward and back

▲ *The layout of controls on a compact disc player. Note the use of a front view and key.*

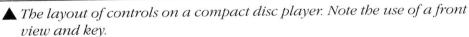

EL TORO RESTAURANT	OPTIMAL KITCHEN MANAGEMENT SYSTEM . DRAFT WORKSHEET 3		
Date: 25/8/95	Operative/s: PABLO		Unit: C3
Menu No: 64 Item: Spanish Rice	**PROGRESS**		
	A.M.	PREP. UNIT	COOK UNIT
INGREDIENTS	**10.00**	Gather all utensils and ingredients to hand	
Item — Quantity			Put water in saucepan
Long grain rice — 150g			Add salt to taste
Onion — 1	**10.05**		Bring to boil
Green pepper — 1		Grate cheese	
Margarine — 25g			
Tomatoes — 395g tin			Add rice to water
Salt — to taste	**10.10**	Deseed pepper and slice. Peel and chop onion. Open can of tomatoes	
Sugar — 1 t/spoon			
Bayleaf — 1			
Cheddar cheese — 50g			
Water — 500 ml	**10.15**		Turn on oven to heat up
UTENSILS			check rice to see if cooked
Item — Size/Qty	**10.20**	Melt margarine in	
Saucepan — 6pt 1			

▲ *Part of a sequence diagram showing stages in the preparation of some food.*

Resource task

CRT 15

...User support

Choosing the best buy

The results of product evaluations can be useful to people who are choosing which product to buy. Meeting a demanding specification can convince people that the product is a good buy.

You will often see specifications for a product used in advertising – especially for cars. It is more helpful for the user to find information about available products in a consumer report. *Which?* and other consumer magazines test and compare products and services, and publish the results with suggested best buys. The consumer knows their advice is independent and objective.

The information in consumer reports needs to be accurate, but also clear and readily understandable to non-technical people.

Charts, diagrams and good graphics help the user to understand the results of any tests, and to decide on the best buy.

4 Designing and making with mechanisms

What can mechanisms do?

Change the type of movement

At the supermarket check-out your shopping is moved forward by a belt and pulley. The motor makes the pulleys *rotate*, but the belt and the shopping on it move in a *straight line*.

You can describe movement in four ways and give each type of movement a symbol.

Linear movement
Movement in a straight line (a car moving along a road, or a ski lift up a mountain).

Rotary movement
Movement in a complete circle (a wheel turning).

Oscillating movement
Movement backwards and forwards in part of a circle (a pendulum).

Reciprocating movement
Movement backwards and forwards in a straight line (a saw sawing).

Change the direction of movement

When you pull *down* on the rope of the hoist, the pulley at the top turns and the load goes *up*.

Resource tasks
MRT 1, 2, 4

...What can mechanisms do?

Alter the axis of rotation

The axis of the gear wheel that drives the hand-whisk is at right angles to the axis of the blades.

This arrangement turns the movement through 90°.

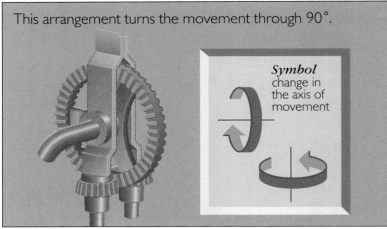

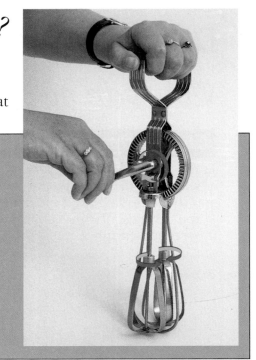

Increase output speed and decrease force

This clockwork mouse uses a gear box to turn the wheels much faster than the drive axle from the motor.

The toy must be light because the moving force provided by the wheels is much less than that applied to the gears by the motor.

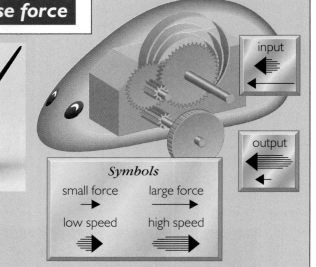

Increase output force and decrease speed

The handles on these pruners act as levers. They are much longer than the blades.

So the cutting force that the blades exert on the branch is much greater than the force applied to the handles.

Resource tasks
MRT 2, 3, 4

Apply and maintain a force

The clothes peg – a spring and two levers – exerts two equal and opposite forces which hold the clothes tightly against the washing line.

Symbol
applied and maintained force →

Transmit movement and force

Mechanisms that transmit force or movement from one place to another are often called **transmission mechanisms**.

In this muck-spreader the linking shaft transmits a turning force from the tractor to the spreader.

The chain on the bicycle transmits rotary movement from the pedals to the rear wheel.

The cables transmit the pulling force on the brake handle to the brake blocks.

Resource tasks
MRT 3, 4

4 DESIGNING AND MAKING WITH MECHANISMS

Using wheels and axles

Wheels are usually *fixed* to **axles** so the wheel and axle turn together.

But sometimes, as in a supermarket trolley, they can spin freely on their axles. Here there are usually bearings between the wheel and the axle to reduce friction (see page 103).

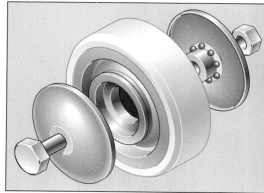

To change the type of movement

The wheels on the wheelchair rotate but the wheelchair moves forward in a straight line. *Rotary* movement has been converted into *linear* movement.

To reduce friction between sliding surfaces

Wheels or rollers are often used to reduce friction. The roller-way enables the lifeboat to be launched easily.

To increase force and decrease speed

The large diameter of the ship's helm means that the helmsman only has to use a small effort to turn the rudder. His hands move a long way to make the rudder move a little.

Using shafts, bearings and couplings

A **shaft** is a rod which transmits rotary movement along its length. There are usually **bearings** between the rotating shaft and its support to reduce friction.

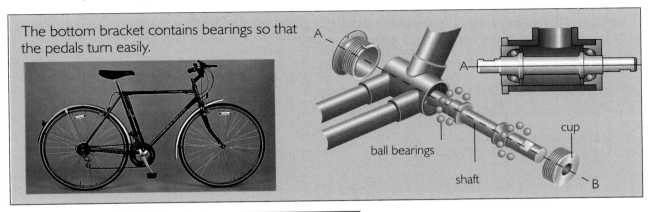

The bottom bracket contains bearings so that the pedals turn easily.

To transmit movement and force

The shaft connecting the pedal crank to the chain wheel in a bicycle transmits movement and force.

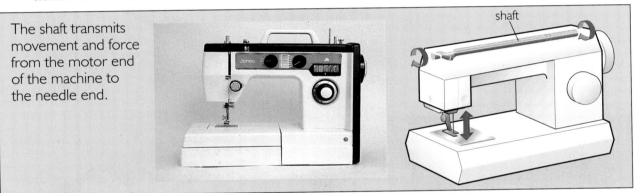

The shaft transmits movement and force from the motor end of the machine to the needle end.

Couplings connect shafts and transmit movement from one shaft to another.

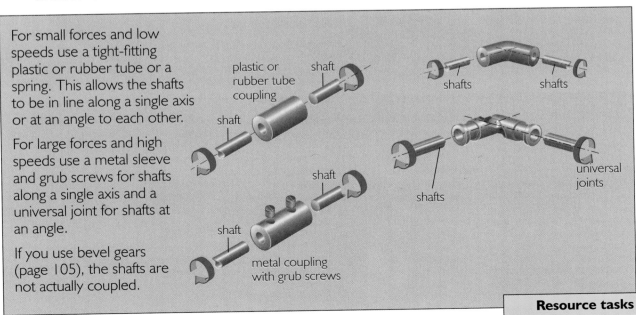

For small forces and low speeds use a tight-fitting plastic or rubber tube or a spring. This allows the shafts to be in line along a single axis or at an angle to each other.

For large forces and high speeds use a metal sleeve and grub screws for shafts along a single axis and a universal joint for shafts at an angle.

If you use bevel gears (page 105), the shafts are not actually coupled.

Resource tasks
MRT 13(6)

Using gears

A gear is a wheel with teeth. It can be fixed to a shaft so that it turns at the same speed as the shaft. You can use one gear wheel (the **input**) to drive another (the **output**) if the two sets of teeth mesh together (a **gear train**).

▲ *Gears come in many different shapes and sizes.*

To change the type of movement

The rack and pinion

This changes *rotary* movement to *linear* movement. The pinion is fixed on a shaft. When the pinion turns it makes the rack move in a straight line. Pulling or pushing the rack makes the pinion turn.

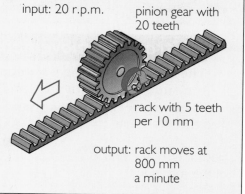

This rack moves 40 mm for every turn of the pinion.

The **input speed** of the pinion is 20 revolutions per minute (**r.p.m.**) so the rack moves at a speed of 20 × 40 mm per minute = 800 mm per minute. This is the **output speed**.

input: 20 r.p.m.

pinion gear with 20 teeth

rack with 5 teeth per 10 mm

output: rack moves at 800 mm a minute

A rack and pinion is used to move the drawer of this CD player in and out. ▼

To change the direction of movement

When two gear wheels mesh the output gear turns in the *opposite* direction to the input gear.

With a third gear wheel in between, the output gear turns in the *same* direction as the input gear. The middle gear is called an **idler gear**.

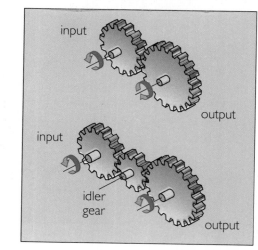

Resource tasks

MRT 13(1), (2)

To change the axis of rotation

Bevel gears
Bevel gears have sloping sides so they can be used for driving shafts that are at an angle to one another.

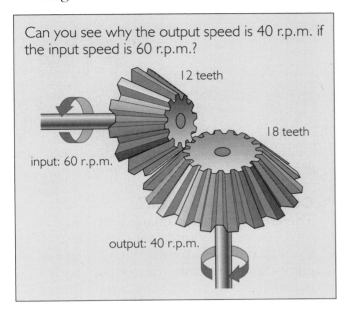

Can you see why the output speed is 40 r.p.m. if the input speed is 60 r.p.m.?

12 teeth
18 teeth
input: 60 r.p.m.
output: 40 r.p.m.

The bevel gears in a hand drill turn the driving force through 90°. Why does the drill bit rotate much faster than the handle?

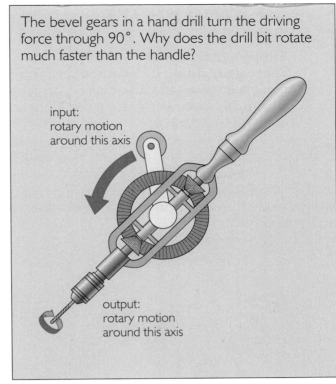

input: rotary motion around this axis

output: rotary motion around this axis

▲ *Turning the driving force through 90°.*

To increase output force and reduce speed

The worm and gear wheel
When this worm gear turns once the gear wheel is 'moved on' by one tooth. The input shaft must turn 22 times to turn the output shaft once. This is **gearing down**.

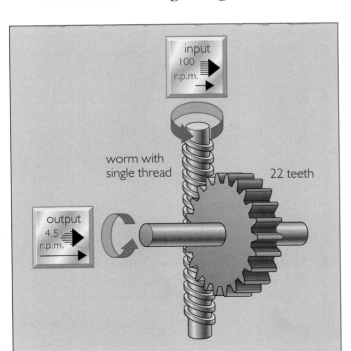

input 100 r.p.m.
worm with single thread
22 teeth
output 4.5 r.p.m.

A worm and gear wheel are used to tighten tennis nets. A small force used to turn the crank handle produces a large force to stretch the net.

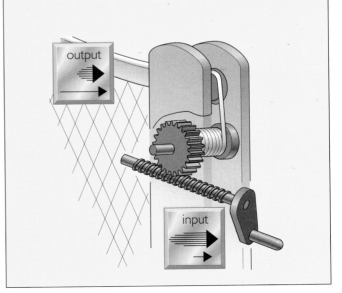

output

input

It is a good way of obtaining large increases in force and large decreases in speed. A *small* force turning the worm is turned into a *large* force turning the gear wheel.

4 DESIGNING AND MAKING WITH MECHANISMS

...Using gears

Simple gear trains
These **gear down** if the output gear has more teeth than the input gear. The output shaft turns more slowly than the input shaft but with more force.

There are three times as many teeth on this output gear as on the input gear. So the input shaft turns three times to make the output shaft turn once. The output shaft turns at one-third the speed of the input shaft, but with greater force.

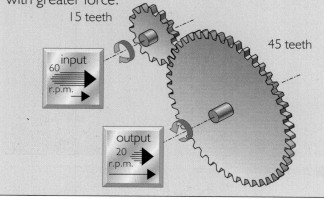

▲ *A simple gear train that gears down.*

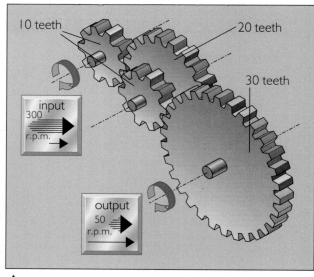

▲ *Gearing down with a compound gear train.*

Compound gear trains
These produce large changes in speed and force. In a compound gear two different-sized gear wheels are fixed to one axle. A compound gear train can contain many pairs of gears.

To decrease output force and increase speed

Simple gear trains
These **gear up** if the output gear has fewer teeth than the input gear. The output shaft turns faster than the input shaft but with less force.

There are three times as many teeth on the input gear as on the output gear. When the input shaft turns once, the output shaft turns three times. The output shaft turns at three times the speed of the input but with less force.

Compound gear trains
Each meshing pair of gears acts in the same way as a simple gear train. When gearing up, the output shaft will turn faster than the input shaft but with less force.

The salad is spun dry by high-speed rotation from a low-speed input.

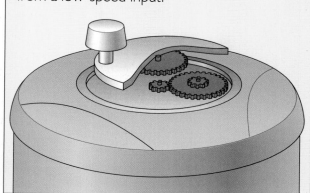

▲ *Salad spinners use compound gear trains to gear up.*

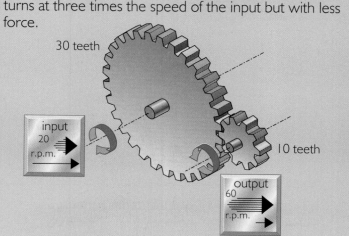

▲ *A simple gear chain that gears up.*

Resource tasks
MRT 5,13(1)

Using pulleys and sprockets

A **pulley** is a wheel used with a **belt** that grips onto it. There may be a groove in the pulley to help the belt grip.

A **sprocket** is a gear wheel used with a **chain** wrapped around it. The teeth on the sprocket fit into the gaps in the chain.

▲ *Belts and pulleys, chains and sprockets.*

To change the type of movement

Belts and pulleys and chains and sprockets change *rotary* movement to *linear* movement. In conveyor belts and escalators a motor provides the input (rotary movement). The output is the object or person moving along in a straight line (linear movement).

To change the direction of movement

You can use pulleys to change the direction of movement.

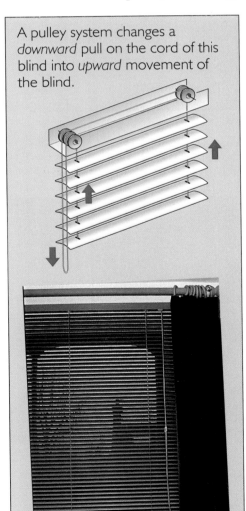

A pulley system changes a *downward* pull on the cord of this blind into *upward* movement of the blind.

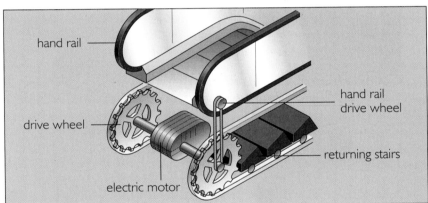

▲ *This escalator is a large chain and sprocket system. The steps are fixed to the chain.*

To alter the axis of rotation

You can alter the axis of rotation by using a twisted belt.

In this vacuum cleaner the axis of the motor is at right angles to the axis of the rotating brush. They are connected by a twisted belt.

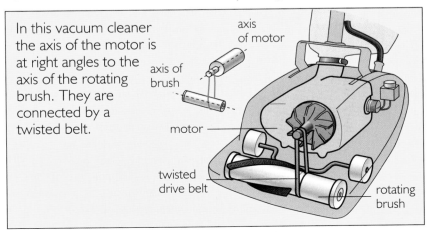

...Using pulleys and sprockets

To increase output force and decrease speed

Using continuous belts or chains

In these systems the sizes of the input and output pulleys or sprockets are important. The output force is increased if the output pulley or sprocket is larger than the input pulley or sprocket. This is the same as 'gearing down' with gear systems.

This washing machine's drum pulley is 15 times bigger in diameter than the motor pulley. The drum turns 15 times more slowly than the motor and the turning force on the drum is 15 times larger than that from the motor.

Using a rope or chain with a loose end

In these systems the number of ropes or chains connected to the output are important. The size of the pulleys or sprockets does not affect the speed or force. These systems increase force if the output force is connected to more ropes or chains than the input force.

Railway engineers use a pulley system to keep overhead cables tight. The input force is provided by iron masses hanging from one rope. The output force holding the overhead cable tight is supplied by three ropes. So it is three times larger than the input force.

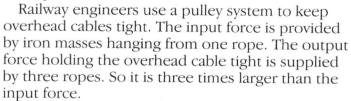
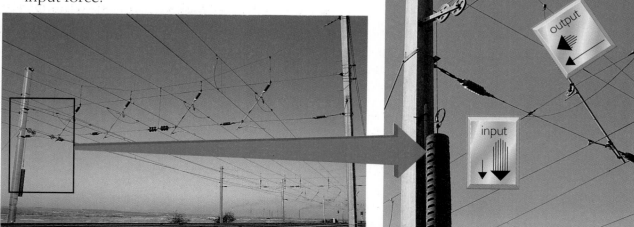

To increase output speed and decrease force

Use continuous belts or chains to do this. The speed is increased if the output pulley or sprocket is smaller than the input pulley or sprocket, as with 'gearing up' with gear trains.

Here the pulley and sprocket make the cutting blades turn faster than the rollers.

The pulley or sprocket connected to the roller (input) is five times larger than that connected to the blades (output). So the blades turn five times more quickly than the roller.

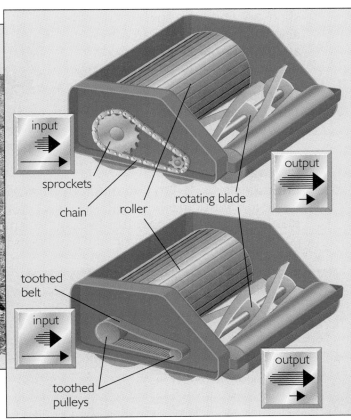

▲ *Drive systems in lawnmowers.*

Positive drive and slip

The drive system of a lawnmower must transmit a large force without slipping. This is called **positive drive**. Both the toothed belt and pulley and the chain and sprocket can do this.

Positive drive is also useful when something needs to be located precisely, such as paper feeding into a computer printer, or when slip would be dangerous, as when riding a bicycle.

Sometimes slip can reduce damage or danger if output is likely to jam occasionally, as with a pillar drill. In these cases belts without teeth are used.

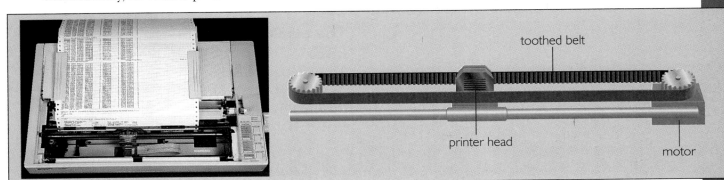

▲ *Precise control of this computer printer head is provided by a toothed belt.*

To transmit movement and force

In the lawnmower and washing machine, movement and force are transmitted with belts and pulleys or chains and sprockets.

Resource tasks
MRT 13 (3), (4)

Using cranks, levers and linkages

Cranks
Cranks are stiff arms fixed to a shaft, so when the shaft rotates, the arm rotates too. They are often attached to handles and used to make turning and tightening easier.

Levers
Levers are bars or rods that move about a **pivot** or **fulcrum**. Sometimes they are made in pairs to grip or crush something.

Linkages
Linkages are used to join different parts together so that they move in a particular way. They are often connected with **pin joints**. You can use **parallel linkages** to make two or more parts move together and/or stay parallel as the linkage moves.

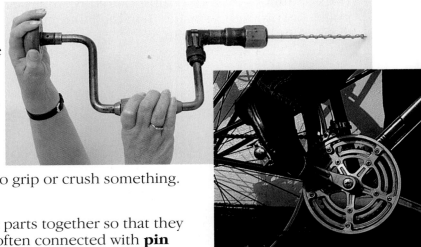

▲ *These cranks make drilling and pedalling easier.*

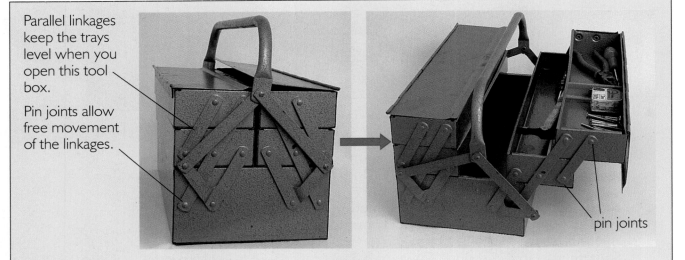

Parallel linkages keep the trays level when you open this tool box.

Pin joints allow free movement of the linkages.

pin joints

To change the type of movement

The small 'crank arm' rotates. It is connected by a link to a lever which oscillates (rocks to and fro). This is often called a **rocking arm**. The fourth link is the base, which does not move.

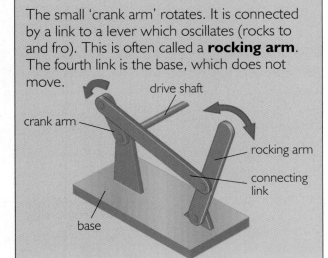

crank arm — drive shaft — rocking arm — connecting link — base

The crank, link and lever
This changes *rotary* movement to *oscillating* movement, called a **four-bar linkage**.

The to-and-fro movement of this electric fan is controlled by a four-bar linkage. ▶

The crank, link and slider

This changes *rotary* movement to *reciprocating* movement (or the reverse). It is similar to the crank, link and lever but has a **slider** instead of a rocking arm. The rotating crank arm makes the slider move backwards and forwards in a straight line.

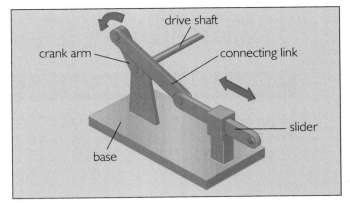

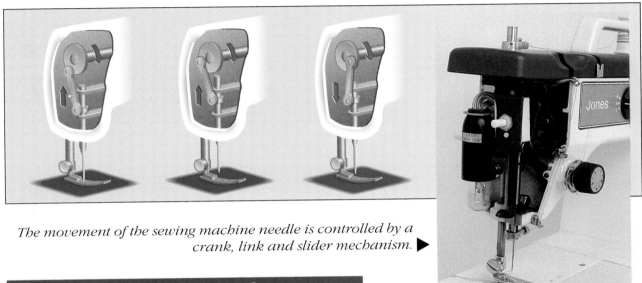

The movement of the sewing machine needle is controlled by a crank, link and slider mechanism. ▶

To change the direction of movement

Reversing direction

You can use levers that have the fulcrum between the input and output to reverse the direction of movement.

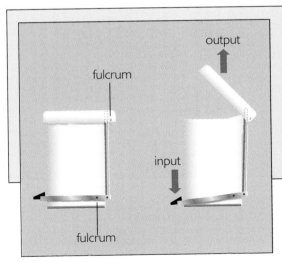

The output movement is opposite to the input movement in this pedal bin. Moving your foot *downwards* makes the lid open *upwards*.

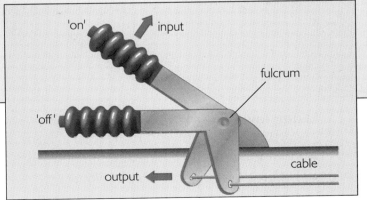

Changing direction through 90°

You can use a right-angled lever called a **bell crank** to change the direction of linear movement.

Resource task

MRT 5

...Using cranks, levers and linkages

To increase output force and decrease speed

Using cranks
The twisting effect of a crank depends on both the force applied (measured in newtons, N) and the length of the crank arm (measured in metres, m). The crank multiplies your effort. The longer the crank arm, the more the effort is multiplied.

The small force you apply to the crank handle becomes a large force on the nut.

Tightening a wheel nut
Twisting effect
= force × distance from axis
= 50 N × 0.2 m
= 10 newton metres (N m).

Using tap handles
To turn the tap on or off needs a twisting effort of 0.25 N m. With the small handle you need to use a force of 10 N. With the long one, only 2.5 N.

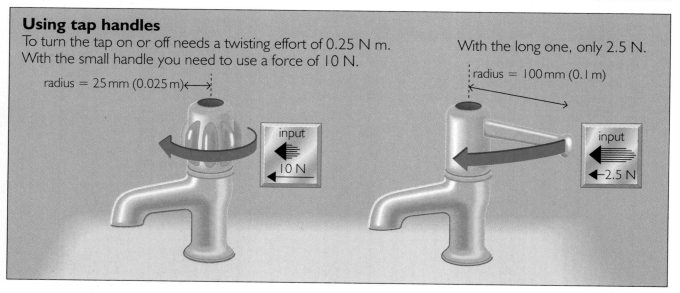

Using levers
The output force you can exert with a lever depends on three things:
- input force (effort), measured in newtons (N);
- distance of the effort from the fulcrum (metres);
- distance of the output force (load) from the fulcrum (metres).

These balance out like this:
input force × distance from fulcrum = output force × distance from fulcrum.

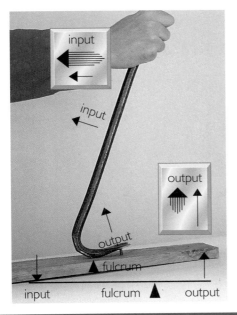

▲ Both these levers increase the output force.

To increase output speed and decrease force

Using levers

In some levers, such as tweezers, you apply the effort between the load and the fulcrum. The tips grip with much less force than the user's fingers, helping prevent damage to delicate objects.

Nail clippers contain this sort of lever. But you would find it impossible to cut nails if this were the only lever. Can you explain why?

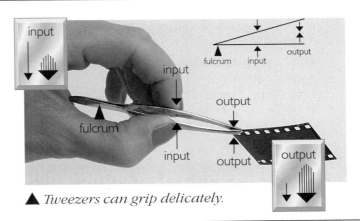

▲ Tweezers can grip delicately.

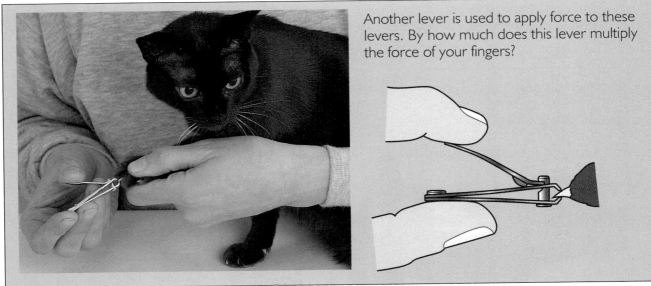

Another lever is used to apply force to these levers. By how much does this lever multiply the force of your fingers?

▲ Nail clippers need to make large forces.

To transmit movement and force

Linkages are often used to transmit movement and force. The linkage in a pedal bin transmits movement and force from the pedal to the lid.

Resource tasks
MRT 7, 13 (5)

Using cams, eccentrics, pegs and slots

All these mechanisms have two parts:
- the input (cam, eccentric, peg);
- the output (follower, slot).

They come in different forms:

- A **cam** is often a non-circular wheel that rotates. As it turns, it pushes a **follower** that moves according to the shape of the cam. The follower is usually a slider or a lever. It is held against the cam by a spring or by gravity.

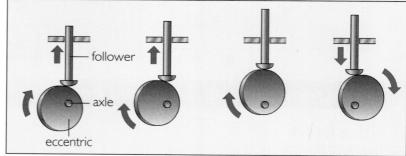

- An **eccentric** is a circular wheel with an off-centre axle. The edge of the wheel can move a follower.

- In a **peg and slot** mechanism, the peg is sometimes fixed to a rotating input. The peg fits into a slot in the follower. As the peg moves in a circle the follower moves backwards and forwards.

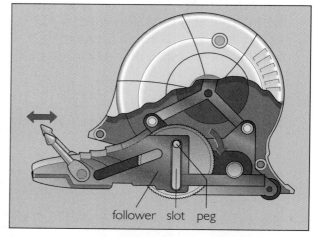

To change the type of movement

Using a cam and follower

Cams change *rotary* movement to *reciprocating* or *oscillating* movement.

The heart-shaped cam (the input) moves the thread-guide of a sewing machine (the output) backwards and forwards along the reel.

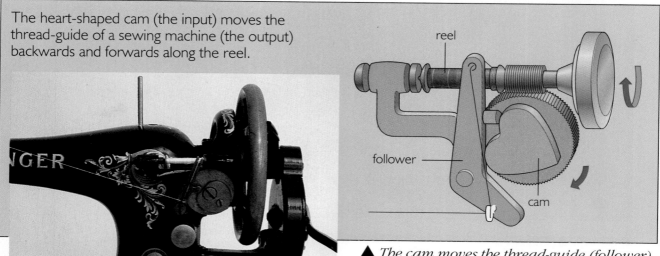

▲ *The cam moves the thread-guide (follower) to ensure that the thread is wound evenly along the reel.*

Resource task

SRT 25

Using an eccentric wheel

You can use this mechanism to change *rotary* to *reciprocating* movement.

This toy fire engine has an eccentric wheel inside fixed to the front axle. When the engine is pulled the eccentric wheel turns, making the driver resting on it bob up and down.

Using a peg and slot

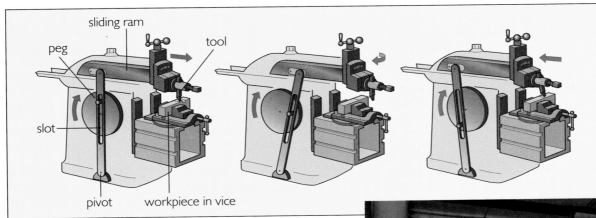

This industrial shaping machine controls a 'quick-return mechanism' with a peg and slot. It turns rotary movement into reciprocating movement. The peg revolving on a wheel moves a slotted lever slowly forwards and quickly backwards. This mechanism is called a crank and slotted lever.

To apply and maintain a force

Using cams

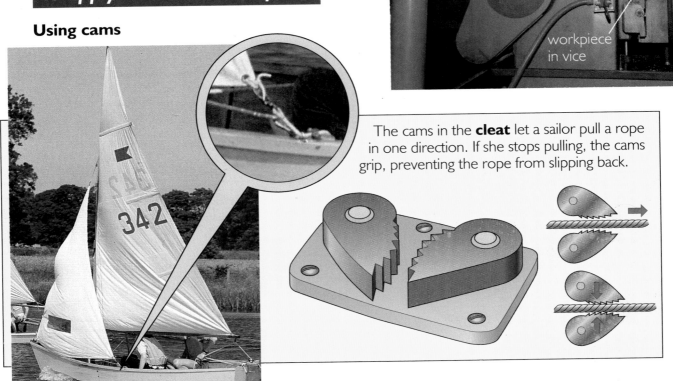

The cams in the **cleat** let a sailor pull a rope in one direction. If she stops pulling, the cams grip, preventing the rope from slipping back.

...Using cams, eccentrics, pegs and slots

Designing cams

Jo had designed a toy band and wanted to use a cam to get a cymbal to go up and down. She wanted the cymbal to move like this:

- rise a lot slowly and then drop suddenly;
- stay still and then rise a little and drop suddenly;
- stay still and then rise a little and drop suddenly;
- stay still and then start the sequence again.

1. Jo wrote this on a sector diagram, adding how far the cymbal had to rise.

2. Then she placed a piece of tracing paper over the diagram and drew a circle to show the smallest part of the cam, with a radius of 30 mm.

3. As the maximum rise of the cam was 20 mm, she drew a concentric circle with a radius of 50 mm.

4. To make the cymbal rise slowly she drew in a curve from the small circle to the large circle over three sectors.

5. To make the cymbal drop suddenly she drew a straight line from the large circle to the small circle along a radius.

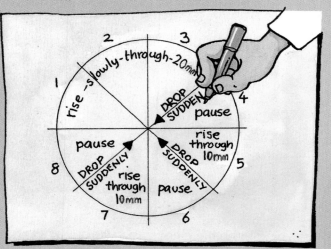

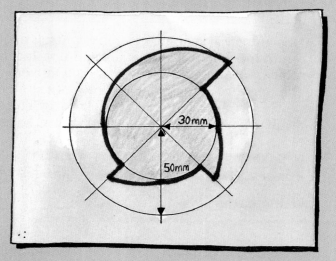

6. Jo completed her shape like this. Can you see how she gets the smaller rises?

7. To make the cymbal stay still she drew a curve along the small circle for one sector.

8. Finally, Jo cut out the cam shape and stuck it onto a piece of plywood. She cut carefully round the shape to produce the cam.

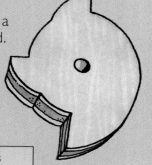

Resource tasks
MRT 10, SRT 25

▲ This is how Jo's cam worked.

Using screw threads

Screw threads come in many shapes and sizes. Which sort you choose will depend on what you want it to do.

The screw thread in this woodworking vice is big: it can be used to apply large forces.

This screw thread is sharp: it can get between wood fibres and grip tightly.

This screw thread is fine: it can be used to level the desk top accurately.

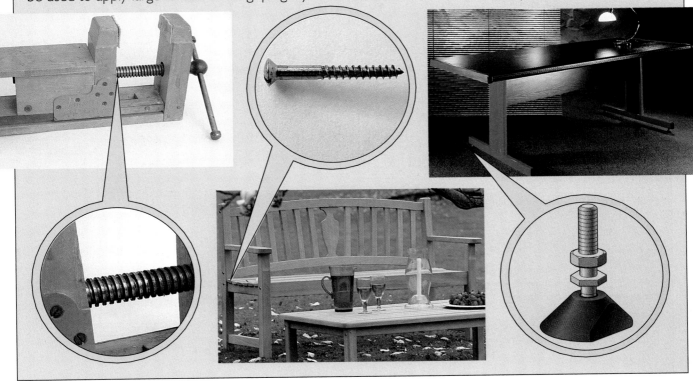

To change the type of movement

The vice and the levelling foot use a screw thread to change *rotary* to *linear* movement. In the spinning top a screw thread is used to turn *linear* into *rotary* motion.

▲ *This screw thread makes the top spin.*

To apply and maintain a force

The screw threads on the jar and lid grip tightly after you have stopped turning the lid.

◄ *Friction keeps the lid on tight.*

Resource task

MRT 9

Mechanisms Chooser Chart

To change the type of movement	You can use:		
From linear to rotating	wheel and axle	rack and pinion	screw thread
	rope and pulley	chain and sprocket	
From rotating to linear	wheel and axle	belt and pulley	screw thread
	rack and pinion	chain and sprocket	
From rotating to reciprocating	crank, link and slider	cam and slide follower	
From rotating to oscillating	crank, link and lever	cam and lever follower	peg and slot
From reciprocating to rotating	crank, link and slider		
From reciprocating to oscillating	wheel and axle	rack and pinion	crank, link and slider
From oscillating to rotating	crank, link and lever	peg and slot	
From oscillating to reciprocating	crank, link and slider	cam and slide follower	

4 DESIGNING AND MAKING WITH MECHANISMS

To change the direction of movement	You can use:		
From clockwise to anticlockwise	gears		belt and pulley
From left to right	levers	linked levers	rope and pulley
From horizontal to vertical	levers	linked levers	rope and pulley
To change the axis of rotation	You can use:		
	bevel gears	flexible couplings	worm and wheel
	belt and pulley		
To increase output force and decrease speed	You can use:		
	With parts rotating or oscillating gears	bevel gears	worm and wheel
	wheel and axle	belt and pulley	chain and sprocket
	With parts reciprocating or moving in a straight line rope and pulley	levers	linked levers
To increase output speed and decrease force	You can use:		
	With parts rotating or oscillating gears	bevel gears	belt and pulley
	chain and sprocket		
	With parts reciprocating or moving in a straight line levers	linked levers	

Using springs

Springs can be used to apply forces and to store and release energy. They come in many shapes and sizes. Which sort you choose will depend on what you want it to do.

To apply and/or maintain a force

This lamp has **tension** springs. When stretched they pull inwards.

Compression springs are in the mattress. When squashed they push outwards.

Torsion springs make the peg work. When opened out they try to close.

Leaf springs provide suspension for this antique car. When bent they try to straighten.

To store energy

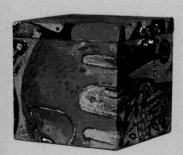

You store energy in the spring of the jack-in-the-box when you close the lid and compress the spring. When you lift the lid the energy is released and the clown jumps up.

Choosing springs

Choose the type of spring that:
- exerts the force you need – push, pull, twist or bend;
- has the stiffness you need – too stiff and it won't change its shape, not stiff enough and it will lose its springiness;
- is the right size to fit the rest of your design.

Remember, compression springs may buckle if they are not supported at the sides or are too long compared with their diameter. Tension springs need a loop or hook at each end to connect to other components.

Resource task
MRT 8

Using syringes

Syringes are useful for making working parts move and for gripping things. You can connect them together with flexible tubing and use them to send force and movement around bends and corners.

They may contain air or water. Air in a syringe will compress if the output movement is resisted. Water will not compress easily but may leak.

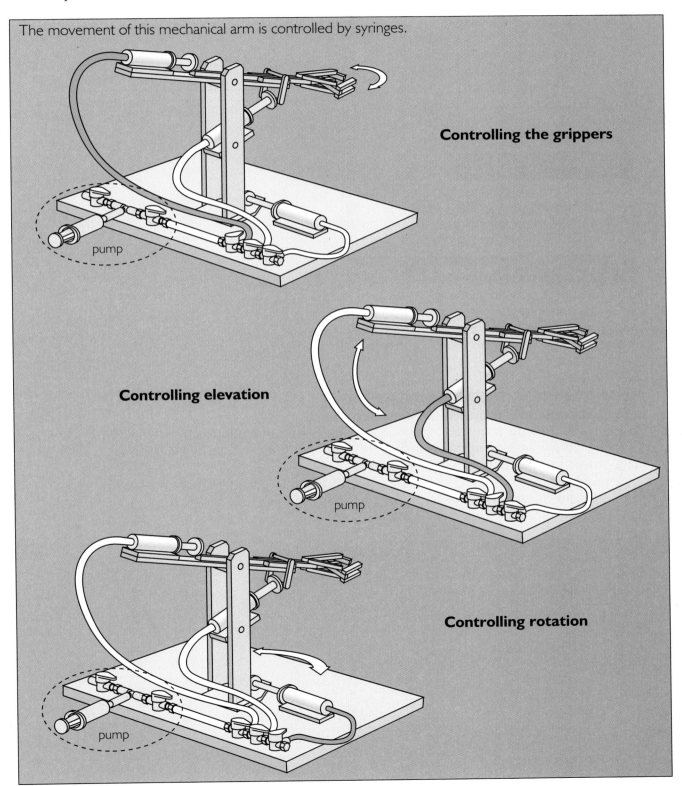

The movement of this mechanical arm is controlled by syringes.

Controlling the grippers

Controlling elevation

Controlling rotation

To change the direction of movement

The tubes connecting the syringes are flexible. They can go round corners, so it is easy to change the direction or angle of linear movement.

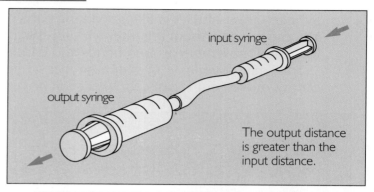

The output distance is greater than the input distance.

To decrease speed

If the output syringe has a bigger diameter than the input syringe, the output distance is less than the input distance. This should also increase force, but friction in the syringe usually prevents it.

To increase speed

If the output syringe has a smaller diameter than the input syringe, the output distance is greater than the input distance.

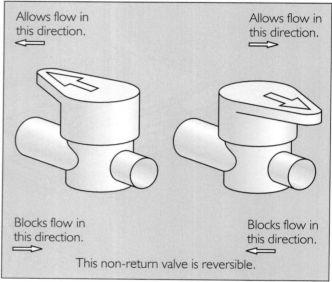

To apply and maintain a force

You can use a non-return valve to let water or air into an output syringe and stop it coming out again. The output syringe will exert a force even if you let go of the input syringe.

▲ *This non-return valve is reversible.*

To transmit movement and force

Syringes filled with water transmit force and movement well. If you try to move a heavy object with an air-filled syringe you will find that the air in the syringe gets compressed instead of the object moving.

Use non-return valves to overcome this. You use the input syringe as a pump and increase the pressure of air in the output syringe so it is harder to compress.

1 Pull out input syringe piston. Air enters input syringe. Non-return valve B prevents air from being drawn from output syringe.

2 Push in input syringe piston. Air is pushed into output syringe. Non-return valve B prevents air leaving output syringe. Non-return valve A prevents air returning to the atmosphere.

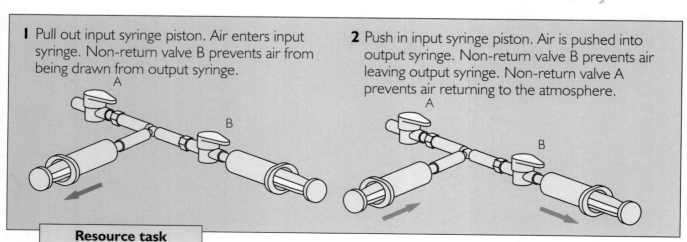

Resource task
MRT 6

5 Designing and making with textiles

Where do textiles come from?

All textiles are made from fibres. The fibres may be natural and come from animals or plants. Or they may be synthetic and come from minerals such as coal or oil.

▲ Fibres from these three sources are twisted to make yarns.

▲ The yarns are made into fabrics in different ways.

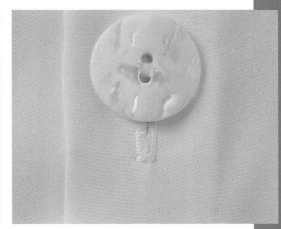

Most fabrics are made by weaving or knitting. **Yarn** for weaving is twisted tightly to make it strong. Yarn for knitting is twisted more loosely to make it stretchy.

What are textiles like?

Words to describe fabrics

To describe a fabric you should be able to say something about its performance, feel and appearance.

Performance		Feel	Appearance
Physical properties	Aftercare		
strong	hard-wearing	warm to touch	plain
flexible	easily cleaned	soft	patterned
stretchy	washable	hard	bright
non-slip	non-iron	smooth	dark
absorbent	quick to dry	rough	shiny
insulating	crease-resistant	cool to touch	dull
flameproof	easy to iron	stiff	coloured
waterproof	stain-resistant	crunchy	neutral
windproof	disposable	scratchy	pastel
lets light or heat or moisture through	biodegradable	fluffy	stripy
	easily dyed	floaty	checked
		furry	flecked
		hairy	see-through

Fabrics Chooser Chart

This chart describes the important properties of different fabrics. Use it to help you choose the fabric that is right for your design.

Fabric	Performance	Feel	Appearance	Cost
Medium calico (undyed cotton)	strong, dyes easily	crunchy	neutral/plain	low
Cotton duck	strong, dyes easily	stiff	neutral/plain	medium
Cotton drill	strong, hard-wearing	quite stiff	plain/coloured	medium
Cotton poplin	absorbent, biodegradable	cool	coloured/patterned	medium
Polyester cotton	crease-resistant, lets moisture through	smooth	coloured/patterned	medium
Cotton T-shirt jersey	absorbent, stretchy	soft	coloured/plain	high
Cotton corduroy	absorbent, hard-wearing	soft, ribbed	coloured/plain	very high
Wool light-weight tweed	insulator	warm, quite hairy	coloured/patterned	high
Wool crepe	insulator, stretchy	soft	coloured/patterned	high
Acetate lining	dries easily, frays easily	silky	coloured/plain	medium
Net	stretchy, non-absorbent	scratchy	coloured/plain, see-through	low
Polyester satin	crease-resistant, frays easily	very silky	coloured/plain	high

Resource tasks
TRT 3, 4, 7, 8, 11–14

Testing fabrics

You can check whether a fabric has the properties you need for your design by testing it. Here are some examples.

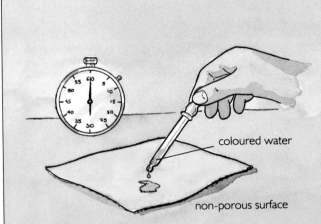

Absorbency
To stay cool, you might need an absorbent fabric that will soak up sweat. Absorbent fabric cleans up liquids as well.

Stain resistance and washability
You might need a stain-resistant fabric. For easy care, a washable fabric will clean and dry better.

Wear resistance
For durability, you might need a wear-resistant fabric that will not go fluffy or bobbly too quickly.

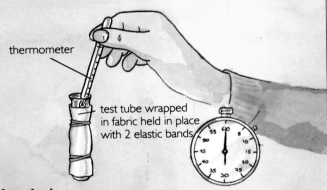

Insulation
To keep warm, you will need a good insulator that will keep warm air in and prevent too rapid cooling.

Windproofing
To keep cold air out, you will need a windproof fabric.

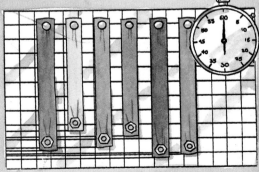

Strength and stretch
For supporting loads or taking strain, you will need a strong fabric. To allow movement, you might choose a fabric that will stretch and then return to its original shape.

Resource tasks
TRT 11–14

Explaining choices

When you design a textile item you should choose a fabric that will perform well, feel good and look right.

The exact properties of a fabric depend on:
- the fibre it is made from;
- the way the yarn has been spun;
- the way the fabric has been constructed – whether it is woven or knitted.

Clothes are made from many different fabrics with different properties. You can see some of the reasons for particular choices below.

Woven fabrics

Woven fabrics do not stretch along the length or width.

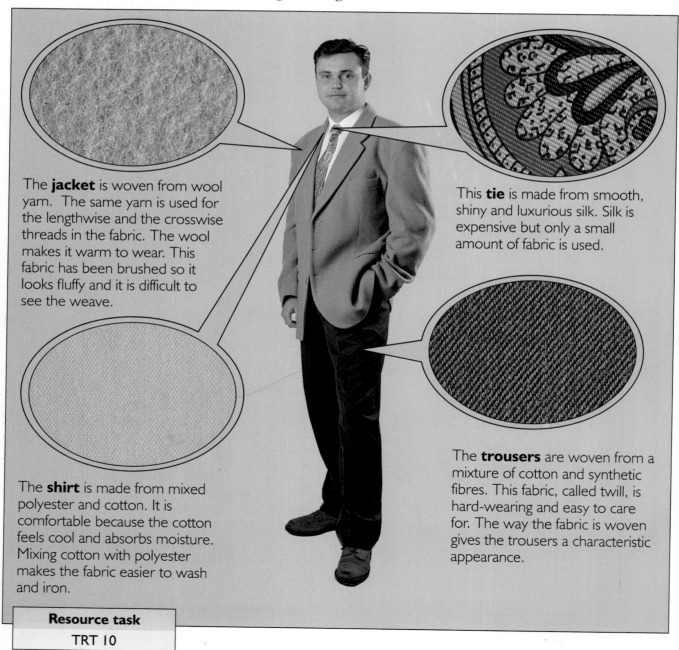

The **jacket** is woven from wool yarn. The same yarn is used for the lengthwise and the crosswise threads in the fabric. The wool makes it warm to wear. This fabric has been brushed so it looks fluffy and it is difficult to see the weave.

This **tie** is made from smooth, shiny and luxurious silk. Silk is expensive but only a small amount of fabric is used.

The **shirt** is made from mixed polyester and cotton. It is comfortable because the cotton feels cool and absorbs moisture. Mixing cotton with polyester makes the fabric easier to wash and iron.

The **trousers** are woven from a mixture of cotton and synthetic fibres. This fabric, called twill, is hard-wearing and easy to care for. The way the fabric is woven gives the trousers a characteristic appearance.

Resource task
TRT 10

Knitted fabrics

Knitted fabrics are stretchy.

This **polo shirt** is made from jersey, the most common knitted fabric. It has a smooth side and a knobbly side and will stretch more sideways than downwards. It is comfortable because the cotton absorbs moisture.

This traditional **Fair Isle waistcoat** is knitted using yarns of different colours. The colour changes make patterns and sometimes pictures.

The **track-suit bottoms** are made from sweat-shirt fabric. This is knitted with extra threads on the inside which are brushed to make the fabric feel soft and warm. The yarn can be all cotton or part synthetic.

Aran knitwear has a stitch pattern that gives a raised effect. The yarn is wool and some of the natural oils are retained. This makes the pullover warmer and shower-proof.

Resource task
TRT 10

Construction techniques – seams

A seam is a line of stitching that joins pieces of fabric together. Seams can be sewn with a sewing machine or by hand. Machines are quicker.

Tips to help you sew successfully

- Mark out the line where you want to sew with tailor's chalk or tacking stitches.
- Use pins or tacking (large running stitches) to hold the pieces of fabric together.

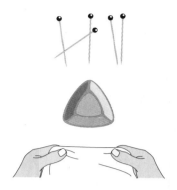

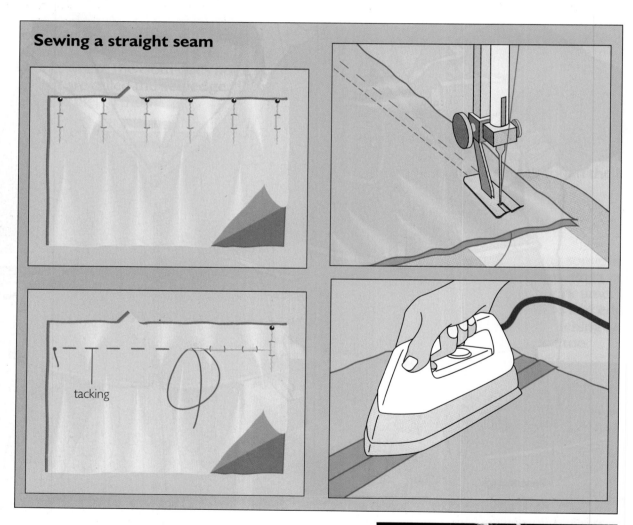

Sewing a straight seam

tacking

Does your fabric fray easily?

If it frays easily you will have to decide how best to stop this happening. See page 132 for ways of finishing edges.

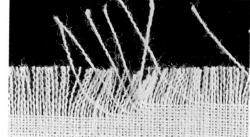

Resource tasks
TRT 1 – 7

Seam allowance

When patterns are cut they allow extra fabric for joining pieces together. This is the seam allowance. Usually it is 15 mm.

If you cut your own patterns you should include seam allowances, and note on the pattern how much you have allowed. If you forget, your design will end up too small.

When you sew the seams, remember to leave the correct seam allowance. Otherwise, your finished product will be too big or too small.

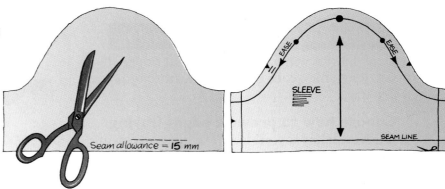

Sewing thread and stitch size

Choose the right sewing thread

To get the best results choose a thread to suit the kind of fabric you are using. Threads come in different thicknesses and fibres. Common ones are polyester and cotton.

Match like with like:
- fine fabric to fine thread;
- thicker fabric to thicker thread;
- synthetic fabric to synthetic thread;
- natural fabric to natural thread.

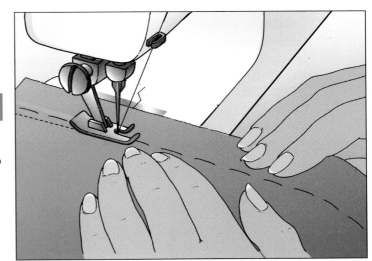

Choose the right stitch size

Most modern sewing machines have a dial to adjust stitch length. Some show a recommended length. It is best to check your stitch length on a spare piece of fabric. Measure the stitches.

As a rough guide, you should use:
- for very fine fabric like acetate lining – 2 mm stitches;
- for medium fabric like poplin or light woollen crepe – 3 mm stitches;
- for thick fabric like cotton duck or corduroy – 4 mm stitches.

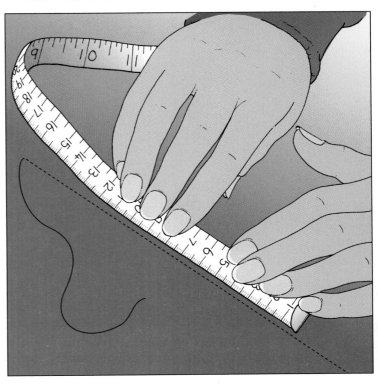

Construction techniques – edges and edging

Preventing edges from fraying

Many fabrics will fray along the cut edges. If you leave the seam edges to fray the item may fall apart.

Choose how to prevent edges fraying

An overlocker is your best way to stop fraying. When you have cut out your pattern, you overlock each piece before joining the parts together.

Preventing stitches undoing

If you are not stitching right up to the edge of the cloth, use a pin and carefully pull both threads through to the wrong side of the fabric. Tie them in a strong knot and snip the ends short.

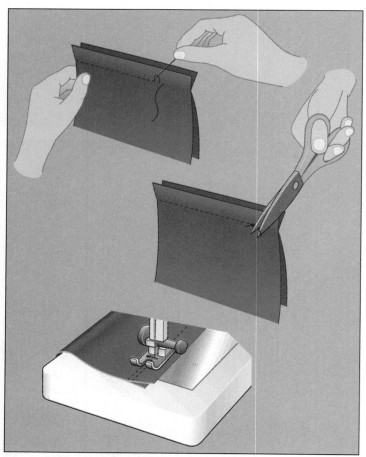

At the beginning of a seam, start 5 mm from the edge, sew a few stitches backwards, then sew the seam. At the end of a seam sew backwards for a few stitches. Cut the ends of thread off.

Most bought garments have the inside edge neatened with stitching called **overlocking**.

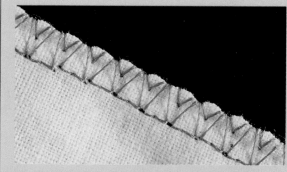

Garments made from knitted jersey fabric, like T-shirts, often have their seams stitched and edged in one operation. This is called **interlocking**.

Zig-zag stitching

If you don't have an overlocker, use the zig-zag stitch on the sewing machine. Stitch near the raw edges of each piece.

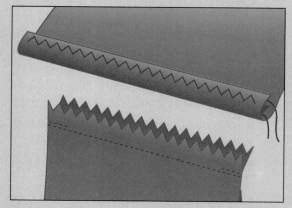

Pinking shears

Special zig-zag scissors, called pinking shears, reduce fraying. With these, you cut along the seam allowance when the seam has been sewn.

Edging techniques

Some edges need to be treated to stop them fraying and make them look neat.

Hemming
One way of neatening an edge is to turn it under and stitch it firmly on the wrong side. This is a hem. You can make hems wide or narrow and sew them by hand or by machine.

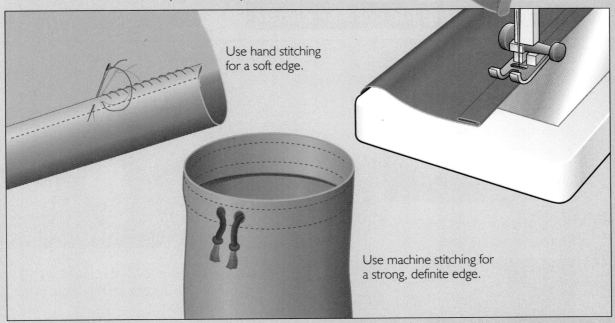

Use hand stitching for a soft edge.

Use machine stitching for a strong, definite edge.

Adding trimmings
You can hide raw edges by adding bindings. These can make the edge more attractive as well as preventing fraying.

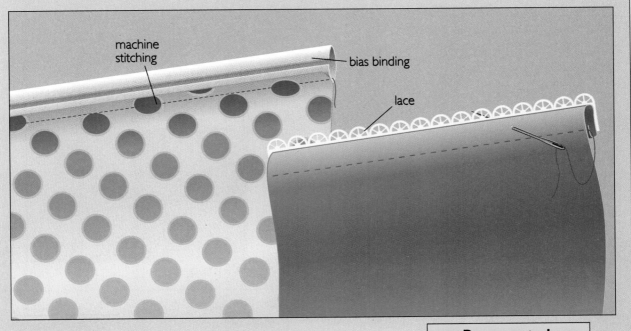

machine stitching

bias binding

lace

Resource task

TRT 6

Construction techniques – cords and elastic

You can use cords or elastic:
- to make garments fit round the waist or cuffs;
- to stop garments falling down;
- to keep a bag closed.

You thread the cord or elastic through a channel called a **casing**.

Making a casing for cord or elastic

1. Turn the top edge of the material under 5 mm and press.

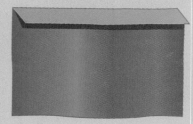

2. Then turn the material under 15 mm and pin in place.

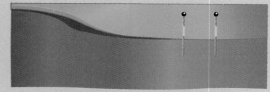

3. Tack the lower edge of the casing in place then machine stitch. Leave a small opening for threading the elastic. Remove the tacks. Machine stitch close to the top fold.

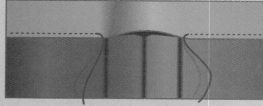

4. Press.

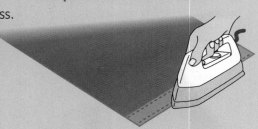

5. Attach a safety pin to each end of the elastic. Pin one to the material as shown. Push the other pin and the elastic trough the casing. Take care not to twist the elastic.

6. Overlap the ends of the elastic by 10 mm and pin. Stitch a square on this area. Stitch across it for strength.

7. Stitch the opening shut. Keep the area flat by stretching the elastic as you sew.

Q 1 How would you change this method if you wanted to use cord rather than elastic?

Resource tasks
TRT 3, 5

Construction techniques – fastenings

Types of fastenings

Fastenings are used to make temporary joins. For example, on a cardigan, buttons allow you to open or close the front; on a rucksack, clips allow you to fasten pockets and straps.

Choosing fastenings

Your choice of fastenings will depend on many factors:

- they look interesting or attractive;
- they are invisible;
- they are wind- or waterproof;
- they can be opened and closed easily or quickly;
- they hold something up;
- they are strong;
- they are long-lasting;
- they are washable or dry-cleanable.

Fastenings Chooser Chart

	Ease of use	Ease of fitting	Variety of types	Strength	Ease of care	Cost
Buttons	●●	●●●●	●●●●	●●	●●	▲▲
		holes ●				
Toggles	●●●	●●●	●	●●	●●	▲▲
Zips	●●●	●	●●	●●●●	●●	▲▲▲
Velcro	●●●●	●●●●	●	●●●●	●●●	▲▲
Hooks/eyes	●	●●	●●		●●●	▲
Press studs	●●●	●●●	●●	●	●●●	▲
Clips/buckles	●●	●●●	●●	●●●	●●●	▲▲▲

● = few blobs for worse, ●●●● = more blobs for better

▲ = cheap, ▲▲▲ = most expensive

Resource tasks
TRT 4, 9

Ways to decorate fabrics

Tie dyeing
Tie small beads or stones into the cloth before it is dyed. The dye cannot go where fabric is tightly tied.

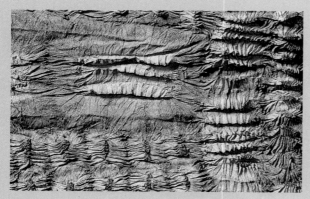

Batik
Trickle or paint wax onto the fabric before dyeing it. The dye cannot go where the wax is. You can make more complex patterns with layers of wax and different dyes.

Transfer painting and direct painting
Brush fabric paints directly onto the fabric. Transfer paints must be painted onto paper. Then place the painted paper over the cloth and iron it. This transfers the pattern onto the cloth, but back to front.

Block printing
Cut or carve a pattern into a block made from lino, potato or wood. Roll ink onto the block and press the block onto the fabric. This transfers the pattern onto the cloth, but back to front.

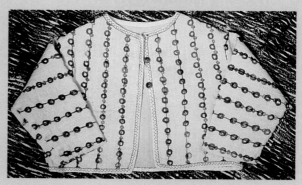

Appliqué
Cut pattern shapes out of different coloured fabrics. Arrange them on the main fabric and sew them down using a close zig-zag stitch and coloured threads.

Embroidery and quilting
In embroidery, you stitch a pattern onto the fabric with coloured silky threads. In quilting, you sandwich a layer of spongy wadding between two layers of fabric. Sew on patterns by hand or with a machine.

Resource tasks
TRT 1, 2, 4 – 7

Choosing the best form of decoration

In deciding which technique to use to decorate your fabric, you should consider all these factors.

Use this Chooser Chart to help you decide on the right technique for your design.

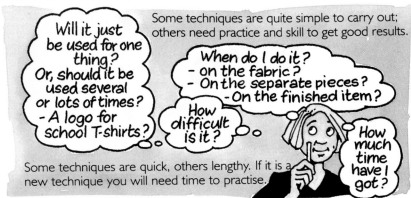

Fabric Decoration Chooser Chart

Technique	One-off or repetitive	Time needed	Simple or complex	When to do it
Tie dyeing	one-off	quick (check time for dyeing and drying)	very simple	before cutting, when in parts, or the finished item
Batik	one-off	varies depending on the detail of motif and number of colours to be dyed	varies, needs practice	before cutting or onto cut parts
Direct painting	one-off	quick for simple motifs, slower for more fiddly ones	simple	onto cut parts or the finished item
Transfer painting	one-off	quick for simple motifs	simple, lettering must be reversed	onto cut parts or finished item
Block printing	repetitive	slow – block must be prepared first, but printing is quick	simple	before cutting
Appliqué	one-off	slow – for careful cutting, positioning and stitching	complex, needs practice	onto cut parts
Embroidery	one-off	varies depending on the detail	simple with practice	the finished item
Quilting	one-off	varies depending on amount and type of stitching	simple with practice	onto cut parts

5 DESIGNING AND MAKING WITH TEXTILES

137

Introducing patterns

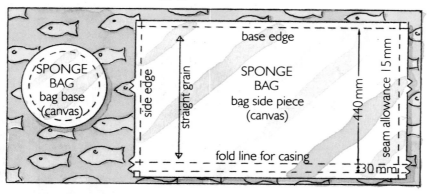

Once you have made notes on the sketch of your design you will need to prepare a paper or card pattern. This is used to mark onto the fabric the exact shape and size of all the pieces you need to make up the design.

Preparing a pattern is called **drafting**. So as not to make mistakes, write all the important information on the pattern.

Use this checklist to make sure that you include all the necessary instructions on your own patterns.

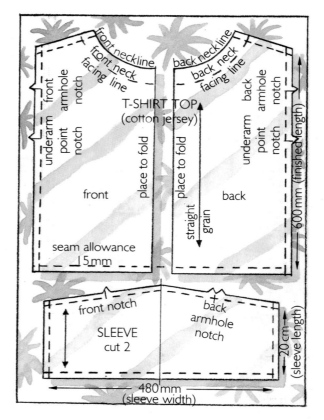

Resource tasks TRT 1–7

How to draft a pattern

1 Checking the design
Your design shows you its shape, its fit, the different parts it should have and where they will be joined together.

Check these details by drawing an exploded view and discuss it with others.

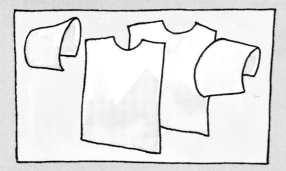

2 Measuring
Now work out the size of each part. Use scrap paper to model your design and estimate sizes. If you are making a garment to fit someone, measure them carefully.

3 Rough draft
Draw each part onto paper as accurately as you can. Add an extra 15 mm all round for seam allowance. Cut the parts out and pin or tape them together with masking tape to make a model. If you are using one of the Resource Task basic patterns, draw them out life size and alter them to fit.

4 Checking the draft
Make any changes needed on the rough paper model. Plan where notches could be used to help with sewing. Work out whether extra fabric will be needed to neaten edges and hems. Note where any extra parts will be attached.

Ask yourself these questions:

Is it the right size? *Is it the right shape?* *Do all the pieces fit together well?* *Does it fit the wearer?*

5 Final draft
Undo your paper model and lay it flat on pattern paper. Use it to draw a neat final draft. Then cut out the pattern pieces.

6 Pattern information
Use the checklist to make sure you have included all the important details.

grain of fabric

7 Planning the layout
Fabric comes in standard widths (900 mm or 1150 mm). Join some sheets of newspaper together to these sizes so that you can plan how to lay out the pieces without wasting fabric. Measure how much fabric you will need.

Designing glove puppets

Puppets, like actors playing a part, usually have a strong character that is obvious from the way they look. They might be a 'goody' or a 'baddy', a 'rich person' or a 'poor person', an 'animal' or a 'person'.

This simple puppet has no character. ▶

Questions to help you design your puppet

1 Starting ideas
Look at existing puppets or pictures of puppets. Books on animation or drawing cartoons will give you ideas on drawing faces with strong expressions. Books on clothes and uniforms will help you with ideas for decoration. You could present the information you collect as an image board.

2 Who is it for?
Is it for you?
If it is for someone else, can you find out what they want?
Can you find out whether they like your suggestions?
Is it a particular character or a type?

3 When or where will it be used?
Is it a toy and used just for fun?
Or is it for a play or show?
Is it going to be used a lot or just once?
Will it have to fit in with other puppets?

4 Shape
What shape should it be?
Will you use the basic pattern or make up your own?
Think about the head, arms and body.
Will it have a face or hair?
What about other details?
How will you get it to look the part?

5 Fit
The opening at the bottom must be big enough to slide in a hand.
The arms and head must be big enough to fit a finger.
How will you check the size?

6 Which fibre and fabric?
Performance – Which properties are important? Which fabrics have these properties?
Appearance – What should the fabric look like? Will you decorate it?
Feel – How should the fabric feel?
Cost – How much can you afford?
Use the Fabrics Chooser Chart on page 124 to help you decide.

▲ *Ideas for puppets.*

Resource task
TRT 1

Designing soft toys

Soft toys should be comforting and appealing. They are usually made of fabrics that feel warm and slightly rough. If they are small they may be fiddly to make. They can be large – big enough to be like a cushion. If they are going to be played with they must be safe and easy to care for.

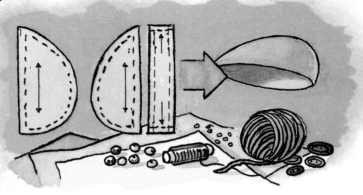

▲ This simple 'animal' shape is easy to make but it does not really look like an animal.

Questions to help you design your soft toy

1 Starting ideas
Look at soft toys or pictures from catalogues. Books on animals will give you ideas for shape. You could present the information you collect as an image board.

2 Who is it for?
Is it for you?
If it is for someone else, can you find out what they want?
Can you find out whether they like your suggestions?

3 When or where will it be used?
Is it for playing with or just for show?
Is it for everyday use or a special occasion?
Will it have to match anything else?

4 Shape
What shape should it be?
Some shapes are easier to make and more interesting than others.

5 Fit
Can you make sure the toy feels right?
You can make it feel firmer or softer by changing the amount of stuffing inside it.

6 Which fibre and fabric?
Performance – Which properties are important? Which fabrics have these properties?
Appearance – What should the fabric look like? Will you decorate it?
Feel – How should the fabric feel?
Cost – How much can you afford?
Use the Fabrics Chooser Chart on page 124 to help you decide.

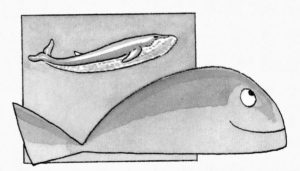

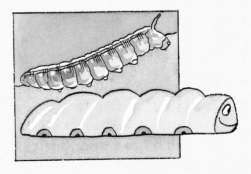

Resource task
TRT 2

Designing bags

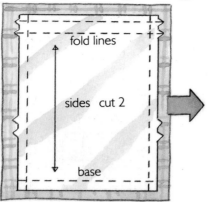
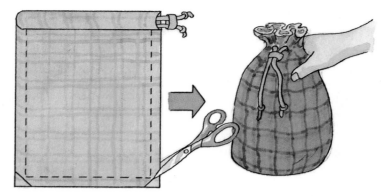

Bags are for carrying things. Some, like pencil cases, are meant to carry a few specific small items, while others, like handbags, carry a wider range of small items. Shopping bags and school bags are designed to carry many things of different shapes and sizes.

Whatever they carry, bags have to be the right size and look good. Many need straps or handles to be useful.

▲ *Can you work out how to change the shape of these simple bags?* ▼

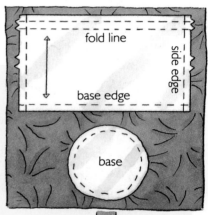

Questions to help you design your bag

1 Starting ideas
Look at existing bags or at pictures from catalogues. You might change a design to suit your requirements. You could produce a style image board to help decide on the fabric or decoration.

2 Who is it for?
Can you find out about what the bag must do and be like?
Is it for you?
If it is for someone else, can you find out what they want?
How will you find out whether they like your suggestions?

3 When or where will it be used?
Is it for everyday or for a special occasion or purpose?
Will it have to match other things?
How will it open and close?
How will it be carried?

4 Shape
What shape and size should it be?
How will you check whether your design is the right shape and size?

5 Fit
What things will be put into it?
If it is designed to contain special things, it must be large enough to fit them in. How can you check this?
How will you check that the person can carry the bag comfortably?
What personal measurements will you need?

6 Which fibre and fabric?
Performance – Which properties are important? Which fabrics have these properties?
Appearance – What should the fabric look like? Will you decorate it?
Feel – What should the fabric feel like?
Cost – How much can you afford?
Use the Fabrics Chooser Chart on page 124 to help you decide.

Resource tasks
TRT 3, 4

Ideas for openings and fastenings

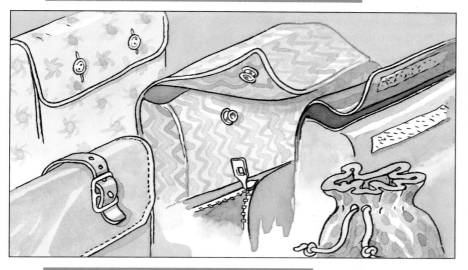

◀ What other ideas can you think of for openings and fastenings?

Making bags stiff

You can make a floppy bag stiff by adding a stiffer material to the structure. You can make the base stiffer with a piece of cardboard or thin sheet of plastic in the bottom (like some rucksacks). You can add wadding to make the sides of your bag both thicker and stiffer.

Ideas for straps and handles

▲ What other ideas can you think of for straps and handles?

Making bags waterproof

Bags may need to be waterproof both on the outside and on the inside.

If the fabric you are using is not waterproof, you could add a waterproofing spray or coating. These are sold in sports and outdoor shops. Or you could add a second layer of fabric that is waterproof, either on the outside or the inside.

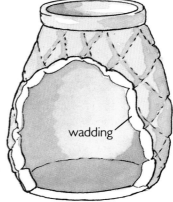

▲ Ways to increase stiffness.

Making the insides look good

Bags need to look good both on the inside and the outside. You need to sew them together very carefully. Use edging techniques (see page 133) to stop the fabric from fraying and to make the seam allowance neat.

Resource tasks
TRT 3, 4

Designing garments from rectangles

The simplest shapes for making garments are squares and rectangles. These have always been combined in different ways all over the world.

Clothes from a single length of cloth

These clothes are made from a single length of cloth with no sewing.

Sarongs from Malaysia

Saris from India and Pakistan

A chamma from Ethiopia

A cloak from Ancient Greece.

A poncho from Peru

A K'sa from North Africa

A sarong from Burma

Squares and rectangles sewn together

Shapes that fit the body more closely can be made from squares and rectangles that are seamed together. Skirts, shorts and tops can all be made from these shapes.

The pictures show the dramatic and beautiful effects of the kimono.

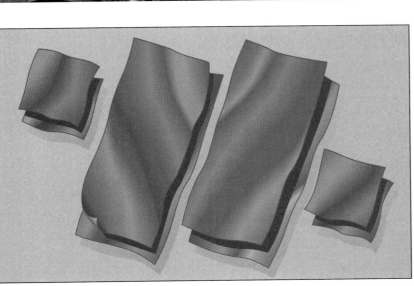

5 DESIGNING AND MAKING WITH TEXTILES

Designing shorts

People wear shorts for sleeping, for beachwear, as underwear, for casual wear. They have to look good, fit comfortably and last well.

▲ *Shorts made using the basic pattern below.*

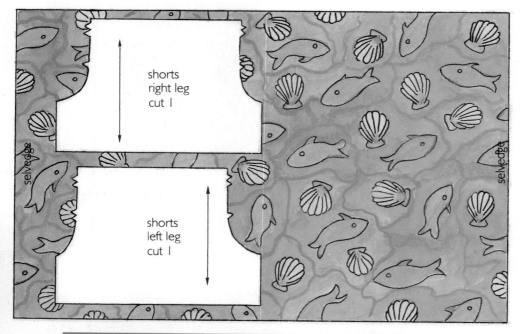

◀ *A basic shorts pattern. There is a Resource Task to help you to alter it to suit your own design ideas.*

Questions to help you design your shorts

Q 1 Starting ideas
Look at existing shorts or pictures from fashion magazines. You could produce a style image board to help decide on the fabric or decoration.

2 Who will wear it?
Is it for you?
If it is for someone else, can you find out what they want?
Can you find out whether they like your suggestions?

3 When or where will they be worn?
Are they for day or night, ordinary or special occasions, sports or leisure?
Do they need to look good with other clothes?

4 Shape
How long should they be?
Are they short, medium length, knee length or longer?
How baggy or tight should they be?
Should they fit close to the body?

5 Fit
Should they have elastic at the waist or a cord to keep them up?

Shorts fit between your legs and round your bottom, so they should be loose enough to allow you to sit and move comfortably.

To check the fit, measure round the hips, at the widest point, and the crotch depth. The basic pattern in the Resource Task fits:
 hips 900mm
 crotch depth 260mm
It can be altered to fit hips and bodies of different sizes.

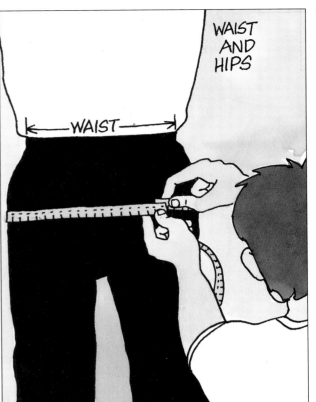

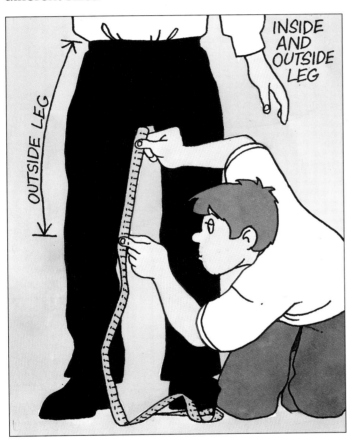

◀ *Taking measurements to check the fit.* ▲
▼

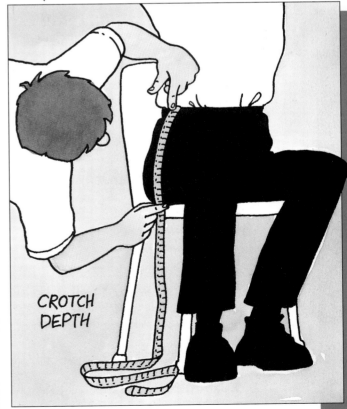

 6 Which fibre and fabric?
Performance – Which properties are important? Which fabrics have these properties? What care will the shorts need after they are made?
Appearance – What should the fabric look like? Should it be plain or patterned? Will you decorate it or will you buy patterned fabric?
Feel – What should the fabric feel like?
Cost – How much can you afford?
Use the Fabrics Chooser Chart on page 124 to help you decide.

Resource task

TRT 5

Designing T-shirt tops

People wear T-shirts for sleeping, as underwear, for casual wear. A T-shirt should look right, fit comfortably and last well.

▲ *This T-shirt style top is made using the basic pattern.*

Questions to help you

1 Starting ideas
Look at existing shorts or pictures from fashion magazines. Try improving on an existing design. You could produce a style image board to help decide on the fabric or fabric decoration.

2 Who will wear it?
Is it for you?
If it is for someone else, can you find out what they want?
Can you find out whether they like your suggestions?

3 When or where will it be worn?
Is it for ordinary wear or special occasions?
Are you designing a top to go with other clothes?
How can you check that the outfit 'works'?

4 Shape
How long should it be? Will the neck opening be round or V-necked? How long should the sleeves be?

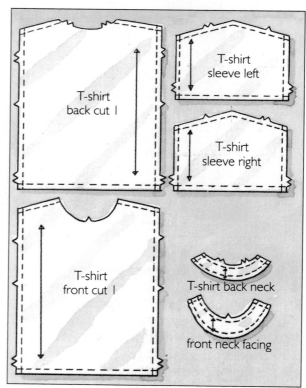

▲ *A basic T-shirt pattern. There is a Resource Task which explains how to draft this pattern and alter it to suit your design idea.*

5 Fit
The neck opening must be big enough to let the wearer's head through. Some fabrics stretch enough for this. How will you check the size?

The T-shirt should be loose enough for you to put it on and take it off comfortably. To check the fit, measure round the chest and the length from the shoulder point. The basic pattern in the Resource Task fits:
 chest 900 mm
 length from back neck to hem 600 mm
 sleeve length 150 mm

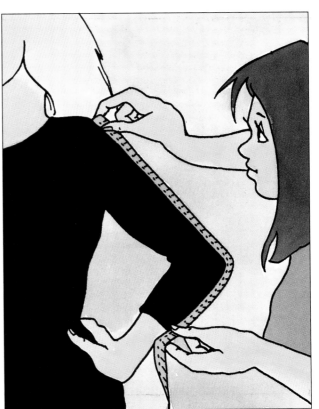

◀ *Taking measurements to check or* ▲
alter the basic pattern. ▼

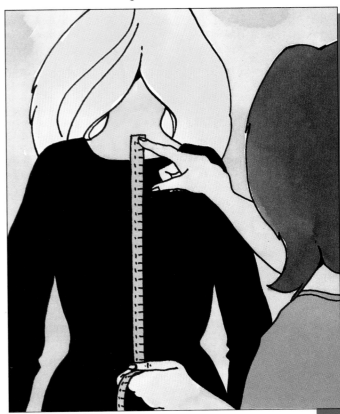

 6 Which fibre and fabric?
Performance – Which properties are important? Which fabrics have these? What care will the T-shirt need after it is made?
Appearance – What should the fabric look like? Should it be plain or patterned? Will you decorate it or buy it as you want it?
Feel – What should the fabric feel like?
Cost – How much can you afford?
Use the Fabrics Chooser Chart on page 124 to help you decide.

Resource task
TRT 6

Designing hats

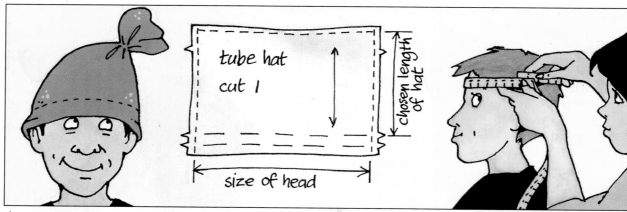

▲ *A simple one-piece hat can be shaped to make it pointed at the top.*

Hats come in many shapes and sizes. They have to fit the head of the person wearing them. Some are adjustable so they can be worn by people with different-sized heads.

They are worn to keep warm, to keep dry, to look good, to make a statement.

Questions to help you design your hat

1 Starting ideas
Look at existing hats or pictures from fashion magazines. Books on fashion history will give you ideas for shape. You could produce a style image board to help decide on the fabric or decoration.

2 Who will wear it?
Is it for you?
If it is for someone else, can you find out what they want?
Can you find out whether they like your suggestions?

3 When or where will it be worn?
Is it for everyday or for special occasions?
Is it to protect the wearer from the weather?
Which is more important – that it does a good job or that it looks good?

4 Shape
What shape and size should it be?
How will you check whether it is the right shape and size?

5 Fit
What measurements will you need to make sure that it fits?
How can you check that it will fit?

6 Which fibre and fabric?
Performance – Which properties are important? Which fabrics have these? Will you need to make the fabric stiffer or waterproof?
Appearance – What should the fabric look like? Will you decorate it?
Feel – What should the fabric feel like?
Cost – How much can you afford?
Use the Fabrics Chooser Chart on page 124 to help you decide.

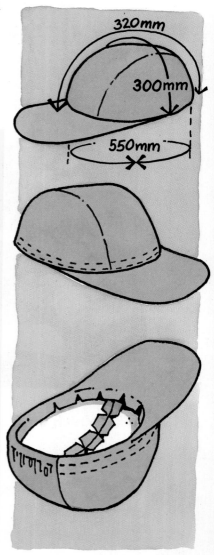

▲ *A baseball type hat can be adjusted to fit any head.*

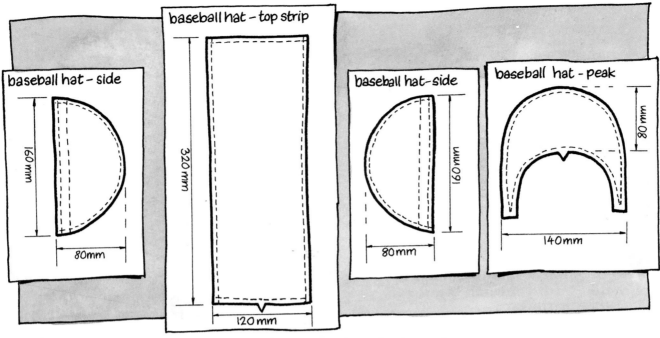

▲ You can work out how to change this pattern to make the peak a different size.

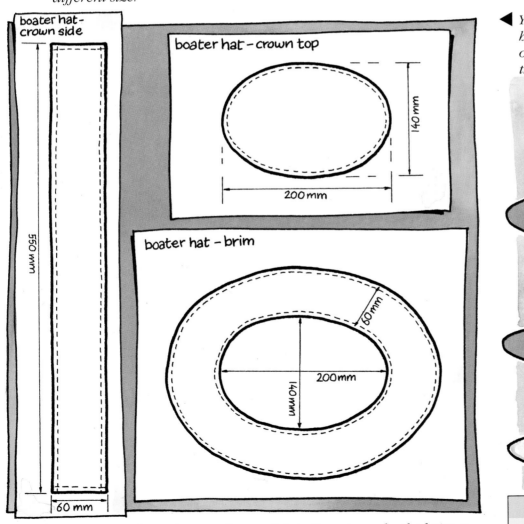

▲ You can work out how to change this pattern to make the brim a different shape.

◀ You can change the height of the crown of this hat as well as the size of the brim.

Resource task

TRT 7

6 Designing and making electric circuits

Products using electricity to make them work all contain circuits to control the electricity. This is usually supplied to the circuit from the mains or from batteries.

Depending on the components in the circuit, you can use electricity to make lights come on, make parts move and make sounds. The size of the electric current can be controlled with resistors. Circuits are turned on and off with switches.

This chapter tells you about these components and how to use them.

Do not connect mains electricity to your circuits. Wash your hands after handling batteries.

circuit symbol

Batteries

Choosing batteries

When you choose a battery, think about:
Size – A large battery holds more energy than a small one. It will last longer but it will cost more. There must be enough room for it in your design.
Voltage – The bigger the voltage, the harder the battery 'pushes' the electricity around the circuit. If the voltage is too large, the components get hot and 'burn out'. If it is too low, they do not work properly. The common voltages available are: 1.5 V, 3 V, 4.5 V, 6 V and 9 V.
Type – Rechargeable batteries are expensive. The three types of non-rechargeable battery shown here are readily available. Your choice will depend on how much current you need to supply and the amount of daily use.

Connecting batteries

▲ *Easy connection is important.*

◀ *Providing 4.5 V from three 1.5 V batteries.*

Batteries can be housed in battery holders, which have metal tags at each end to connect them to the circuit. With more than one battery in a holder you must make sure they are the right way round.

Some batteries can be connected to a circuit with press-stud fittings. The connector shown for the PP3 battery will only fit one way round, so you cannot connect it the wrong way.

Resource task
ECRT 1

Making connections

Join components together in a circuit with plastic-coated copper wire. Remove the plastic insulation where the join is made. Use solder or special connecting components for the join.

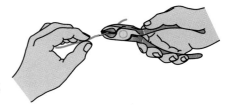

Using solder

Soldering is quick and effective. Good soldering is essential: a badly soldered joint can stop a circuit from working and make fault-finding difficult. It is easy if you follow these rules. **Take care not to rest the hot tip of the soldering iron on the electric cable. Make sure that there is plenty of fresh air where you are working.**

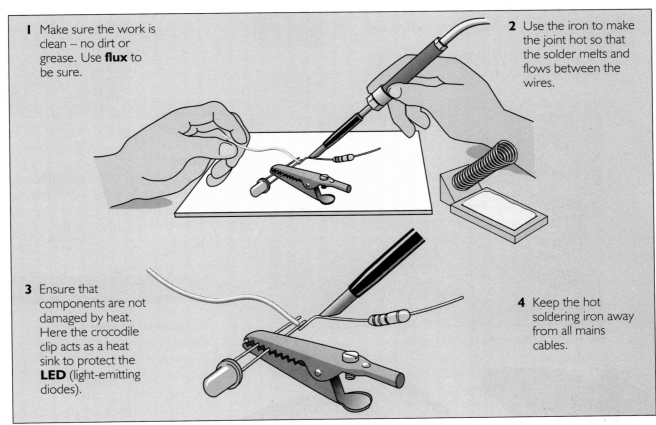

1. Make sure the work is clean – no dirt or grease. Use **flux** to be sure.

2. Use the iron to make the joint hot so that the solder melts and flows between the wires.

3. Ensure that components are not damaged by heat. Here the crocodile clip acts as a heat sink to protect the **LED** (light-emitting diodes).

4. Keep the hot soldering iron away from all mains cables.

Using connectors

You can also make connections using crocodile clips, jack-plugs and connection blocks. These are useful for making temporary connections when you are testing a circuit to see if it works.

You can make temporary circuits on breadboards for testing before you make a more permanent circuit using soldered joints. Batteries often have press-stud connectors.

Resource tasks
ECRT 2, 3

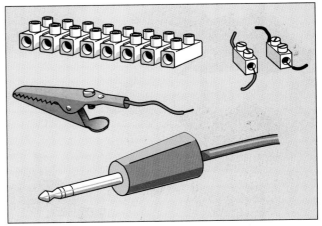

▲ *Other ways of connecting components.*

Components to make things happen

Making light

Light-emitting diodes (LEDs)

LEDs light up when current passes through them. They are often used as **indicators** in control panels because they do not get hot and only use a small current.

In the circuit shown here a **resistor** is being used to limit the current passing through the LED. This stops it being damaged by too large a current.

You can also get LEDs that flash, and do not need a protecting resistor.

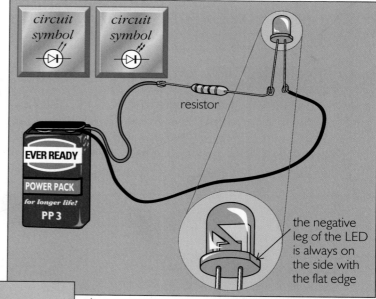

▲ It is important to connect LEDs properly.

Light bulbs

Light bulbs light up when a current passes through a very thin wire – the filament – of the bulb. The filament gets hot and glows brightly. Use light bulbs for lighting rather than indicating.

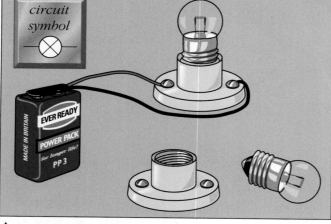

▲ Small bulbs screw into holders.

Making movement

Electric motors

When an electric current is fed to a motor, the shaft of the motor rotates. You can use this shaft to drive moving parts.

Electric motors range in size from very small to very large. The shafts of small motors often rotate quickly, but produce little useful twisting effect.

You can use gears to decrease the rotational speed and increase the twisting effect.

◄ The electric motor inside a cassette player is very small.

Making sounds

Use bells and buzzers as sounds for alarm and warning systems. Buzzers must be connected the right way round.

Protecting components

Resistors control the size of an electric current. They restrict the flow of a current by providing a **resistance** to it.

The value of the resistance is measured in ohms (Ω).

You can use resistors to protect other components in your circuit, like LEDs and transistors (see pages 163–6), from too large an electric current.

The resistor colour code

Four coloured bands on the resistor show the value of the resistance provided. The first three bands give the value; the fourth, usually silver or gold, gives the accuracy to which it has been made (the tolerance).

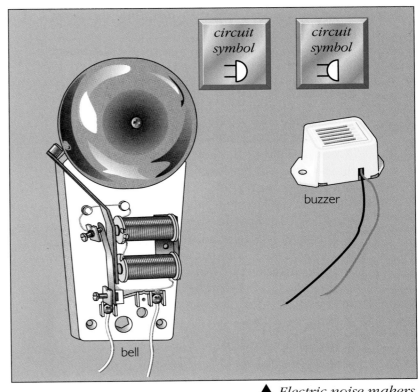

▲ *Electric noise makers.*

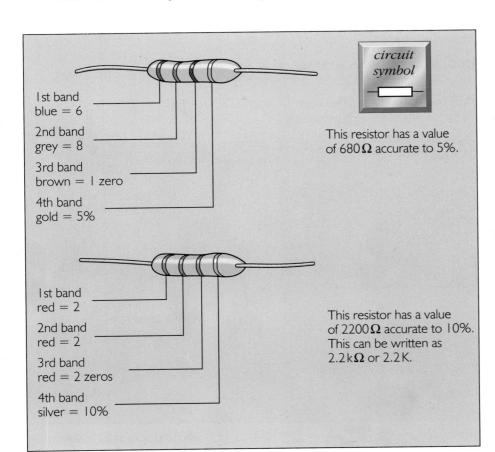

1st band blue = 6
2nd band grey = 8
3rd band brown = 1 zero
4th band gold = 5%

This resistor has a value of 680 Ω accurate to 5%.

1st band red = 2
2nd band red = 2
3rd band red = 2 zeros
4th band silver = 10%

This resistor has a value of 2200 Ω accurate to 10%. This can be written as 2.2 kΩ or 2.2 K.

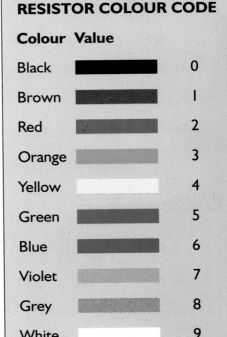

RESISTOR COLOUR CODE

Colour	Value
Black	0
Brown	1
Red	2
Orange	3
Yellow	4
Green	5
Blue	6
Violet	7
Grey	8
White	9

◀ *Calculating resistor values.*

Resource task
ECRT 7

Using switches

Switches turn a circuit on or off by 'making' or 'breaking' connections. There are several types of switch.

For setting on or off

An **on/off switch**, such as a light switch stays in the off position until you switch it on. It stays on until you switch it off.

Simple on/off switches are called **single-pole, single-throw** switches. The **poles** are the number of circuits that the switch makes or breaks. The **throws** are the number of positions to which each pole can be switched.

There are four types of on/off switches:

- **push switches** are pushed on and off;
- **toggle switches** are flicked on or off;
- **slide switches** are pushed or pulled;
- **rocker switches** are usually pressed.

For holding on or off

Two types are available:

- the **push-to-make switch**, which *makes* a connection when pushed;
- the **push-to-break switch**, which *breaks* a connection when pushed.

Both return to their original position when you stop pushing. You find push-to-break switches on car doors and refrigerators.

Whichever type of on/off switch you choose in this circuit the light remains off until you switch it on. Then it stays on until you switch it off.

▲ *On/off switches.*

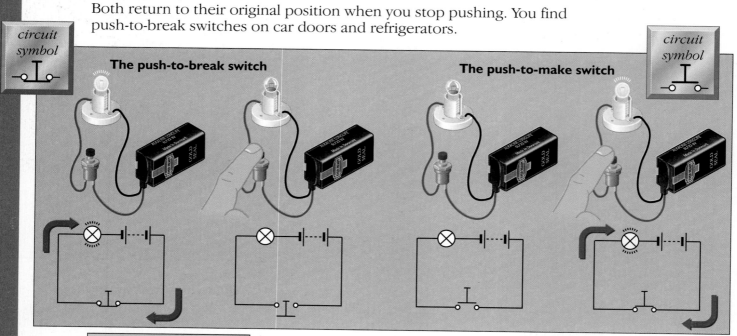

▲ *Switches you hold on or off.*

Resource tasks
ECRT 4, 5

For turning on and off at the same time

These are called **change-over switches**. The circuit diagrams show a **single-pole, double-throw** change-over switch turning off one light bulb while turning on another. This could be a slide, toggle or rocker switch, or a **micro-switch**, which needs only a very light touch to operate.

A **double-pole, double-throw** change-over switch, which could be a slide, toggle or rocker type, is used to reverse the direction of an electric motor. They are used for the rewind in cassette tape players.

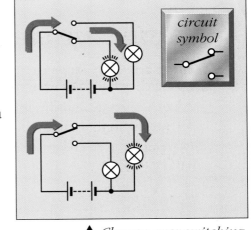

▲ *Change-over switching.*

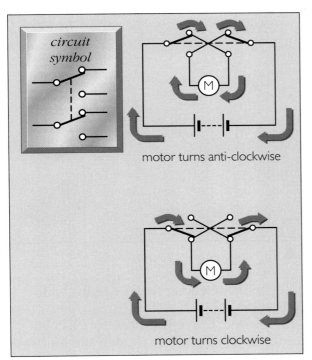

▲ *Reversing a motor.*

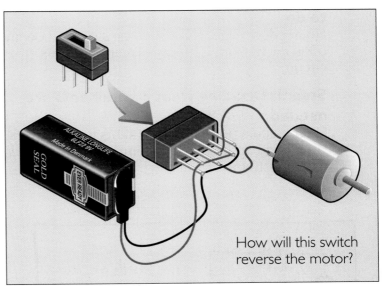

▲ *Wiring a reversing switch.*

Special switches

Tilt switches

These contain a blob of mercury that moves when the switch is tilted. Electricity can flow through mercury, so when the blob covers both connections, the circuit is made and electricity flows.

Tilt switches are useful in security systems to detect things being moved.

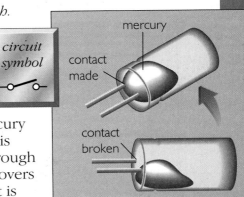

▲ *The tilt switch.*

Reed switches

These are controlled by a magnet. When the magnet is close to the switch the contacts are held together and the circuit is made, allowing current to flow. When the magnet is removed the contacts separate and the circuit is broken.

Reed switches are useful in security systems to detect intruders opening doors or windows.

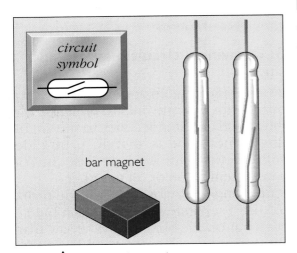

▲ *The reed switch.*

Resource tasks
ECRT 4, 5

6 DESIGNING AND MAKING ELECTRIC CIRCUITS

Designing electric circuits

There are three steps to designing an electric circuit:
1 Decide what the circuit needs to do – the specification.
2 Decide what components are needed.
3 Draw a circuit diagram and try it out.

Two pupils used these steps to design electric circuits for part of a design and make task.

Pradip needed to put lights in a model house

Step 1: Deciding what the circuit needs to do
He needed a light that could be switched on and off in each room – the specification for his circuit.

Step 2: Deciding what components are needed
He needed a power source, lights and switches. He looked at the Electric Components Chooser Chart for help.

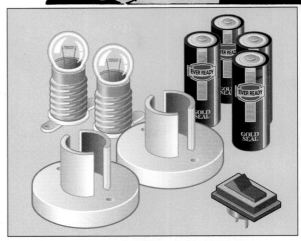

▲ *The components Pradip needed.*

For the lights he chose two 6 V bulbs with bulb holders.
For the power source he chose four 1.5 V batteries providing 6 V (to match the bulbs) of zinc chloride type for medium current and regular use.
For the switches he chose a small rocker switch to set the lights on or off.

Step 3: Drawing a circuit diagram and trying it out
He drew a circuit diagram and made it up on a breadboard with just one bulb to try it out. When he added another in series to the circuit the bulbs did not shine as brightly. As he added extra bulbs they all gave less light. Also, he could only switch them all on or off together.

By connecting the bulbs in parallel he realized he could use a separate switch for each bulb. Trying this out on the breadboard he saw that the bulbs did not get dimmer when more than one was switched on. He drew the circuit diagram for this new circuit.

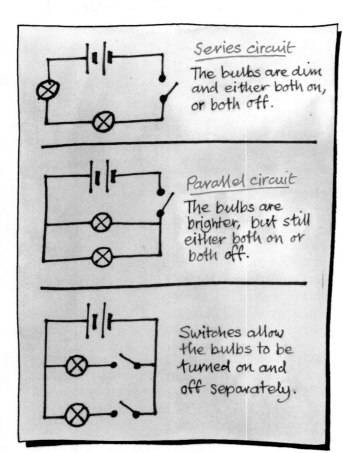

▲ *How Pradip's ideas changed.*

Resource tasks
ECRT 6, 7

Sally needed to power a model buggy

Step 1: Deciding what the circuit needs to do
She wanted the buggy to go forwards and backwards. This was the specification for her circuit.

Step 2: Deciding what components are needed
She needed a power source and a means to make the wheels go round in either direction. She used the Electric Components Chooser Chart to help her choose the right components:
- for rotational movement she chose a small electric motor that worked on 6–12 V;
- for the power source she chose a 9 V battery (to match the motor) of alkaline type for high current and heavy use;
- for a way to reverse the current she chose a double-pole, double-throw switch, in the form of a toggle switch because this is easy to operate.

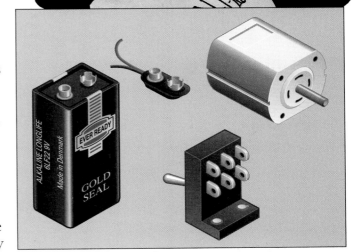

▲ Sally's components.

Step 3: Drawing a circuit diagram and trying it out
Sally drew the circuit diagram for a reversing circuit (see page 157) and made up the circuit using leads and crocodile clips. A friend suggested that an improvement would be to add a warning signal when the buggy went backwards.

She used the Chooser Chart to decide on a buzzer. She drew the new circuit diagram and tried it out. It worked, so now she could make the permanent circuit on the buggy with soldered connections.

Can you see any problem with the new circuit?

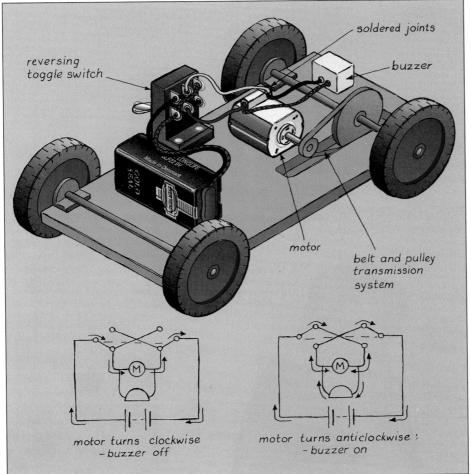

▲ The completed buggy with a warning circuit.

Electric Components Chooser Chart

What the component might need to do	Options	Symbols	Points to check
To provide a power supply	batteries: • zinc carbon for low current, infrequent use • zinc chloride for medium current, regular use • alkaline for high current, heavy use	+ − ‖- - -‖	Make sure voltage of battery is suitable for components in the circuit.
To make light			
To give a signal	a light-emitting diode		Use protecting resistor. Must be correct way round.
	a flashing light-emitting diode		Does not need protecting resistor. Must be correct way round.
To provide illumination	a light bulb	⊗	Must match power source.
To give rotary movement	an electric motor	Ⓜ	Must match power source. May need 'gearing'.
To make sound	a bell		Must match power source.
	a buzzer		Buzzer must be correct way round.

What to do if your circuit doesn't work

Use this checklist before you ask your teacher.

Must have got something wrong... where's that diagram?

Check carefully against your circuit diagram.

Maybe the battery isn't connected properly?

Check to be sure.

Might be a dud battery...

Test it with a light bulb that you know works.

Perhaps it's the solder?

Check for any 'dry' solder joints.

Could be a loose connection somewhere...

Look carefully to check.

Must be something else not working...

Remove components one at a time and test them in a circuit that does work.

What the component might need to do	Options	Symbols	Points to check
To control current size			
By setting at a fixed value	a fixed resistor	—▭—	Value provides the required current.
To switch			
To hold something on or off	a push-to-make switch		Two connections to the switch.
	a push-to-break switch		
	a reed switch		
	a tilt switch		
To set something on or off	a single-pole, single-throw switch		Two connections to the switch. Which type will be most suitable for the user?
	● push switch		
	● slide switch		
	● toggle switch		
	● rocker switch		
To turn something on and something else off	a single-pole, double-throw change-over switch		Three connections to the switch. Which type will be most suitable for the user?
	● micro-switch		
	● slide switch		
	● toggle switch		
	● rocker switch		
To reverse direction	a double-pole, double-throw change-over switch		Four connections to the switch. Which type will be most suitable for the user?
	● slide switch		
	● toggle switch		
	● rocker switch		

7 Designing and making electronic circuits

Electronic control

You can think about electronic control by using the systems thinking method described in Chapter 2, *Strategies*.

Electronic control devices have **inputs** and **outputs** so are often called **electronic control systems**. The indicator system on a car is an example.

By operating switches (the input) electric current flows through circuits (the processor), causing lights to flash in a particular way (the output).

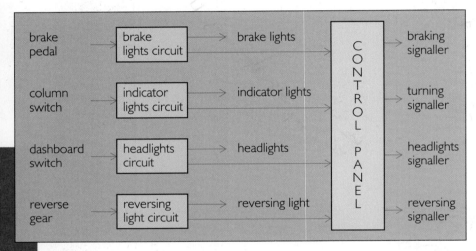

Depending on the input, the circuit will process the information so that the left or the right indicator flashes, or both flash as a hazard warning.

The indicator lighting system is a subsystem of the whole car lighting system, which includes brake lights, reversing light, headlights and control panel lighting.

The input to an electronic control system is provided by sensors which react to changes in the surroundings, such as temperature, light level, noise level or moisture.

Electronic control in action

162

Sensors and processors

Sensors

Sensors can tell you when their surroundings have changed. The devices shown below all contain electronic control systems using sensors.

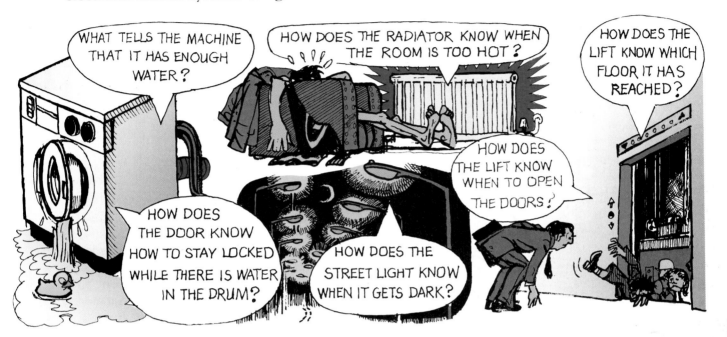

Types of sensor

Light-dependent resistor (LDR)
Use this sensor to detect changes in light level. It has a high resistance in the dark and a low resistance in the light.

Moisture sensor
Use this sensor to detect rain water or tap water. It can be designed as a small printed circuit board or two probes. Water droplets bridge the gap between the tracks or probes, letting a small electric current flow.

Thermistor
Use this sensor to detect changes in temperature. Usually it has a high resistance at low temperatures and a low resistance at higher temperatures.

Bimetallic strip
Use this sensor to detect changes in temperature. The **bimetallic** strip changes shape by curling up as it gets hot.

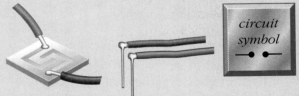

Resource task
ECRT 10

...Sensors and processors

Transistors

These are used to process the information from sensors. They come in many shapes and sizes but all have three legs. Each leg has a special name – the **emitter**, **base** and **collector**.

Their positions tell which leg is which. This is important, because if you connect a leg wrongly the circuit will not work and the transistor will be damaged.

Note the positions of the legs on the circuit symbol for a transistor.

▲ *Different types of transistors.*

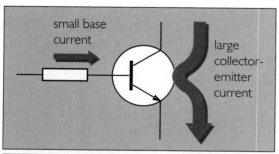

What they do

A transistor is like a switch with no moving parts. It is switched on when a small current flows into the base leg. Then a large current can flow through the transistor from the collector to the emitter. When the current into the base leg stops, the transistor turns off.

As the small current has caused a larger current to flow, transistors are sometimes called **amplifiers**. A transistor is usually protected from too high a current by a resistor on the base leg.

Setting up electronic circuits

Note how the components are connected between a positive and negative rail in electronic circuits. In this way it is easy to connect many different components into a circuit.

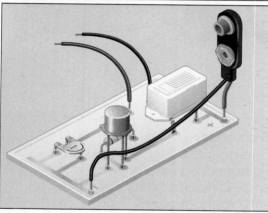

Variable resistors

These allow you to vary the size of the current in a circuit by changing the value of the resistance. You can use them with sensors to alter the sensitivity of a sensing system.

Unlike fixed resistors, they have three legs. Usually you connect the middle and one of the outside legs to your circuit.

Variable resistors come in different shapes and sizes. You adjust the resistance by turning a control with your fingers or with a small screwdriver.

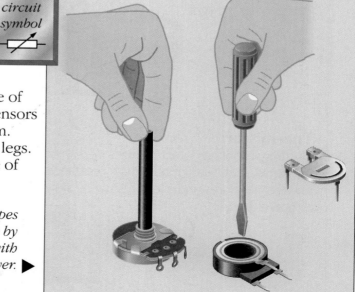

Resource task
ECRT 9

Using sensors with transistors

A lighting control system

This electronic control system detects when the surroundings become dark and lights a bulb. Look at the panel to find out how it works.

In the light, a small electric current flows from the positive rail through the variable resistor and the LDR (which has a low resistance in the light) into the negative rail. No current flows into the base of the transistor so it stays switched off. So no current can flow through the bulb that is connected to the transistor.

In the dark, a small electric current flows from the positive rail through the variable resistor but cannot flow through the LDR as this now has a high resistance. The current is channelled into the base of the transistor which turns the transistor on.

A larger current now flows through the transistor and the bulb lights up. When it becomes light the resistance of the LDR goes down. The small electric current is no longer directed to the base of the transistor so the transistor turns off and the light bulb goes out.

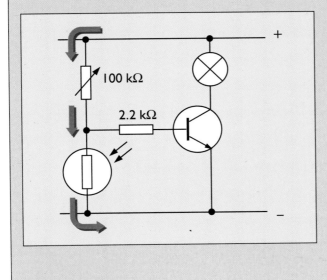

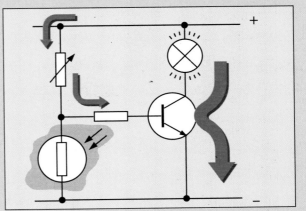

On a printed circuit board this circuit looks like this:

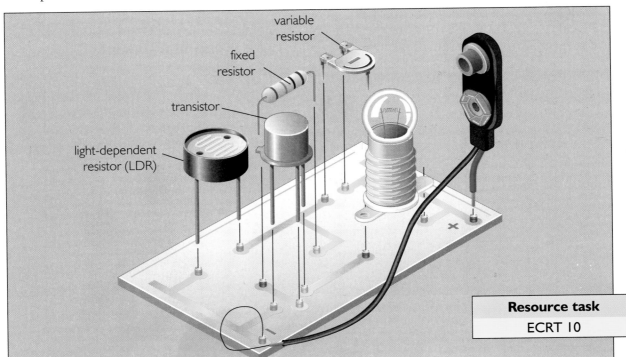

Resource task
ECRT 10

7 DESIGNING AND MAKING ELECTRONIC CIRCUITS

165

...Sensors and processors

Sensing arrangements

Here are some useful sensing arrangements.
These two-resistor arrangements are called **potential dividers**.

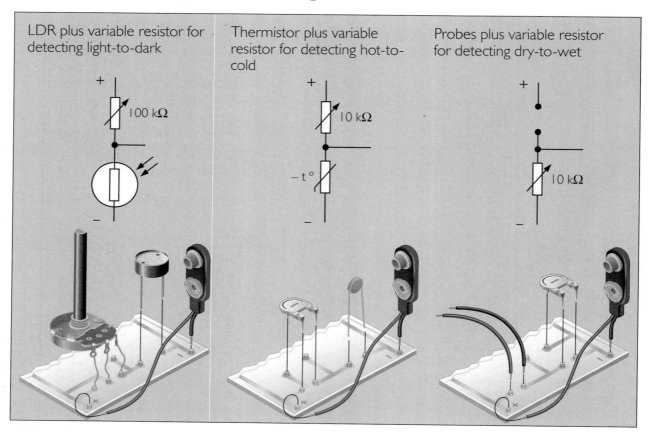

Using transistors in pairs

If you use two transistors with a sensor you can get a more sensitive and faster-operating control system. This arrangement is called a **Darlington pair**.

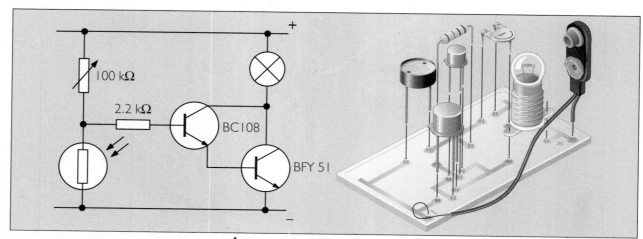

▲ *The circuit diagram and printed circuit board arrangement for detecting changes in light level, using a Darlington pair.*

Resource task
ECRT 11

The relay

What it does

The current output from a transistor may not be large enough to drive the output device – an electric motor, perhaps. Here you can use a **relay**.

A relay is made of two parts – an electromagnet and a set of switches. When a small current passes through the electromagnet it is turned on and operates the switches. These are connected to another circuit with a more powerful supply of electricity, which is used to drive the high-current output device.

Protecting the circuit

Always use a **diode** with a relay to prevent the electromagnet from damaging the rest of the circuit.

This is the circuit board arrangement to connect a relay into a circuit.

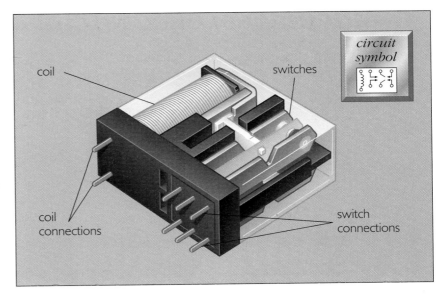

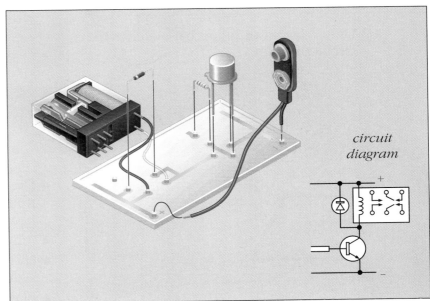

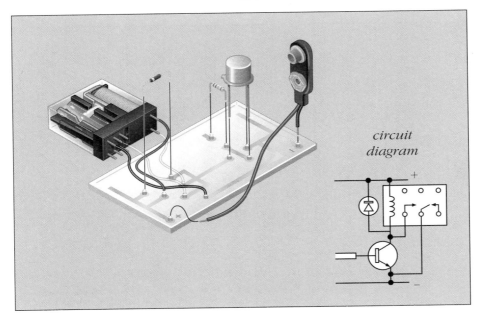

Latching

You can design the circuit so that the output stays on even when the input signal has stopped.

When the relay operates, the switches close and bypass the transistor, so the relay stays on even if the transistor turns off. This is called using a **latch**.

Resource task
ECRT 12

Making a printed circuit board (PCB)

1 Start by drawing the circuit diagram.

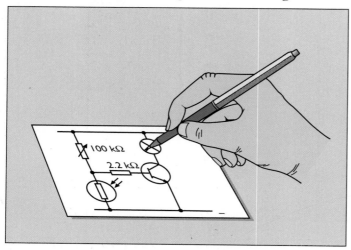

2 Draw a layout diagram that matches the circuit diagram.

3 Draw out a mask of the layout on acetate sheeting. Use dry transfers or a computer and plotter for this. Your teacher may give you a ready-produced mask.

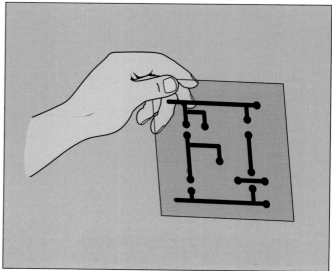

4 Use a **UV** light box and the mask to transfer your layout onto photosensitive board.

- remove the black plastic from the board;
- lay the board face down on top of the acetate mask;
- place in the UV light box.

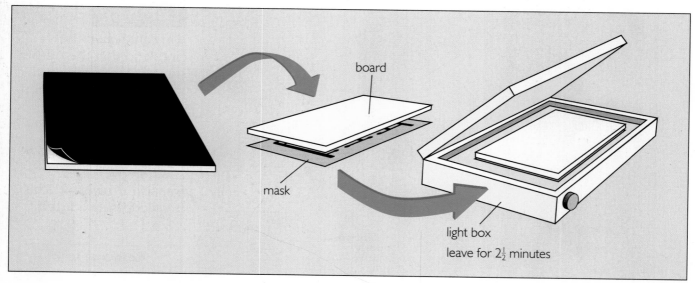

5 Develop the layout with sodium hydroxide solution (**take care – this is toxic and causes burns – wear eye protection and gloves**). Rinse the board under the tap.

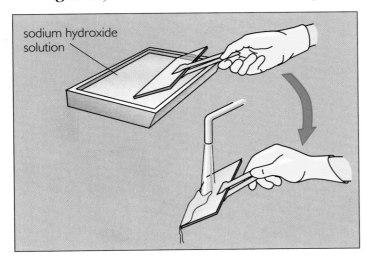

6 Etch in ferric chloride solution. (**Take care – this is toxic and causes burns and stains – wear eye protection and gloves. Make sure that there is plenty of fresh air in the area where you are working.**)

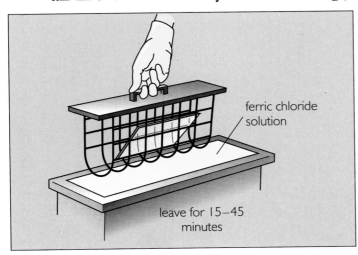

7 Wash the board under the tap and dry it. Clean the track with a PCB eraser.

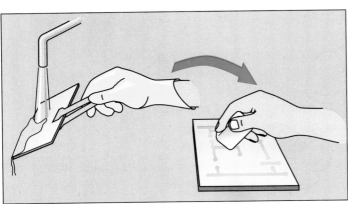

8 Drill holes for **components**.

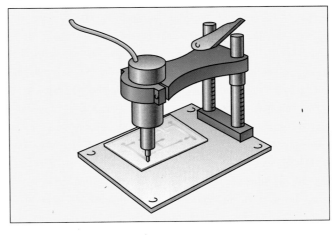

9 Insert components and solder them into place.

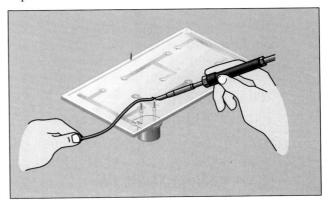

10 The completed circuit.

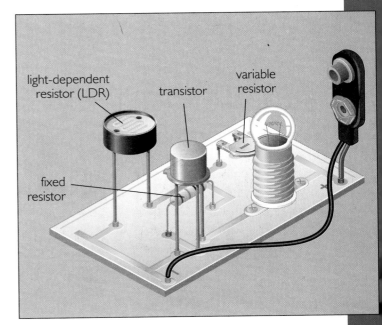

Designing an electronic product

Points to consider

In designing an electronic product you need to think about:
- the electronic system – the input, processor and output;
- the product casing and style;
- the user interface.

The Sensing with Electronics Chooser Chart on page 172 will help you.

Emma designs an overfill alarm

Emma's gran can't see well enough to fill cups, kettles and saucepans to the right level. Emma decided to design an electronic product that would tell her gran when containers were full of water. Emma developed this specification:

What it has to do:
- detect when liquid has reached the level;
- show that this level has been reached;
- be usable with a range of containers.

What it should look like:
- not be easily noticed (Emma's gran didn't want it to be too obvious).

Other requirements:
- it should be easy for an elderly person to operate;
- it should be easy to keep clean.

Emma used the questions in the Sensing with Electronics Chooser Chart to work out the details.

The electronic system
For the **input**:

What does it need to detect? .... water level.

What sensors can I use? ... moisture sensor.

For the **processor**:

Will the signal from the sensor need to be increased? .... tap water... poor conductor... so, yes, a single transistor should do it.

For the **output**:

What does it need to do? ... make a noise ... a buzzer... not too loud.

Is it a high-current device? no, so I won't need a relay.

Does the output need to stay on after the input has stopped? ... no, gran will stop pouring as soon as she hears the buzzer.

Resource task

ECRT 10

Emma drew a block diagram of the system and noted the components needed for each block, and their symbols.

She turned the block diagram and circuit symbols into a circuit diagram. From this she worked out the PCB layout diagram.

```
inputs  →  processor  →  output

water probe    one or two       buzzer
               transistors

10 kΩ variable   2.2 kΩ fixed
   resistor       resistor
```

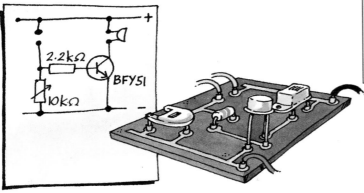

▲ *Emma's final circuit.*

Emma found some tubing that would be easy to hold. She made up several different shaped card developments that seemed stable and good to look at without being too noticeable.

The user interface
She thought through how to make it easy for gran to use.

- What controls will gran need?
- Will it need an on–off switch?
- Will gran need to test that it is working OK?
- How will gran know which way up to hold it?

The product casing and style
Emma then thought about how her gran would use the device.

By answering these questions and using the modelling strategies suggested in the Sensing with Electronics Chooser Chart she was able to design a layout that met all her gran's requirements.

Resource task
ECRT 10

7 DESIGNING AND MAKING ELECTRONIC CIRCUITS

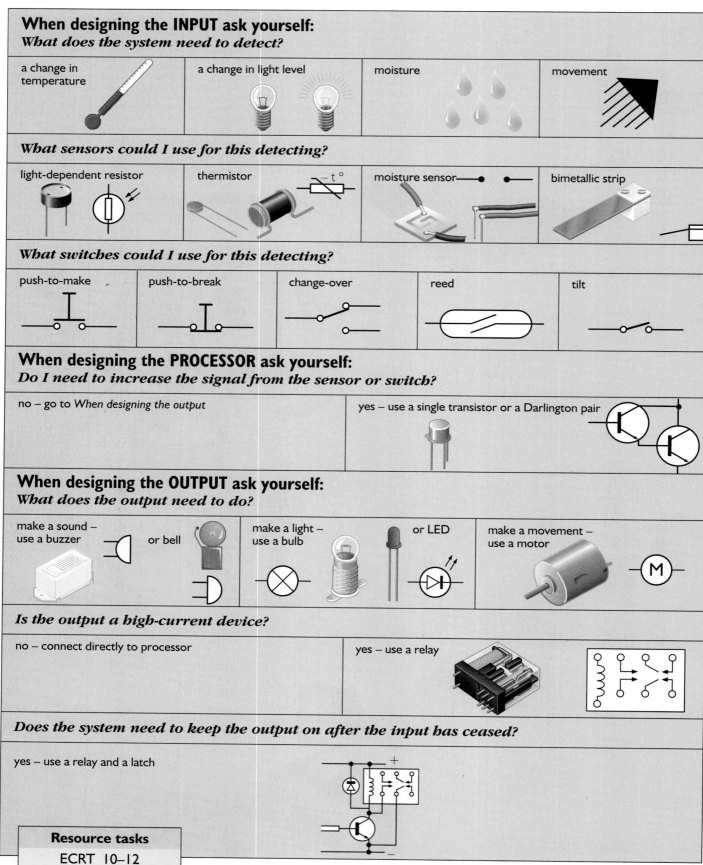

7 DESIGNING AND MAKING ELECTRONIC CIRCUITS

When designing the PRODUCT CASING AND STYLE ask yourself:

What overall shape and size would be suitable?	large enough to take the contents but small enough for hand-held use – try modelling for fit
What does the user like?	try using an image board
How can I get an appearance that fits in with where it will be used?	try out some ideas against the image board

When designing the USER INTERFACE ask yourself:

What switches, other controls or indicator lights will the user need?	try an imaginary user trip
How can I make the layout of the controls look easy to understand?	try modelling with a plan
How can I make it clear what each switch or indicator light is for?	try labelling with signs or symbols
How can I position switches and other controls so they are easy to operate?	try ergonomic modelling

What if it doesn't work?

Ask yourself these questions:

Is the battery working and the right way round?

Check to be sure.

Is everything in the right place?

Check against your layout diagram.

Are all the components the right way round?

Check the transistors, LEDs and diodes.

Are there any loose connections?

Look carefully to check.

Are there any dry joints?

Check carefully.

Are there any cracks in the copper tracks of the PCB?

Look carefully to check.

8 Computer control

Most modern electrical machines are controlled by a complex integrated circuit called a **microprocessor**. It is programmed to work the various electrical devices in the machine.

The car wash

The microprocessor in this car wash controls:
- valves for the spray-jets;
- electric motors for the roller and wheel-scrubbers;
- hot-air blowers to dry the car;
- hydraulic cylinders to move the roller and driers into position.

Information from sensors makes sure that the machine works safely and well.

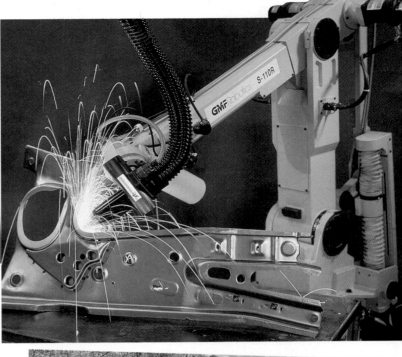

▲ The motorist presses a push switch to start the wash sequence. Sensors are used to position the roller and wheel-scrubbers correctly for the size of car being cleaned.

▲ You can use a computer and interface box to design and work the same kind of control system.

Computer control systems

Think of a computer control system like the diagram below.

The signals processed inside a computer are very low voltage and cannot be used directly to power electrical devices.

An interface box has its own electrical power supply. It is used to drive the output devices connected to it. Circuits inside the box allow the tiny signals from the computer to switch this power on and off.

Input signals from switches and sensors are also processed in the interface box before they pass into the computer.

The two kinds of input to a computer control system are:

- **digital inputs** which come from switches that are either on or off.
- **analogue inputs** which come from sensors that sense constantly changing values such as temperature, light or sound levels.

Analogue input signals must be converted into digital form before they can be processed by the computer.

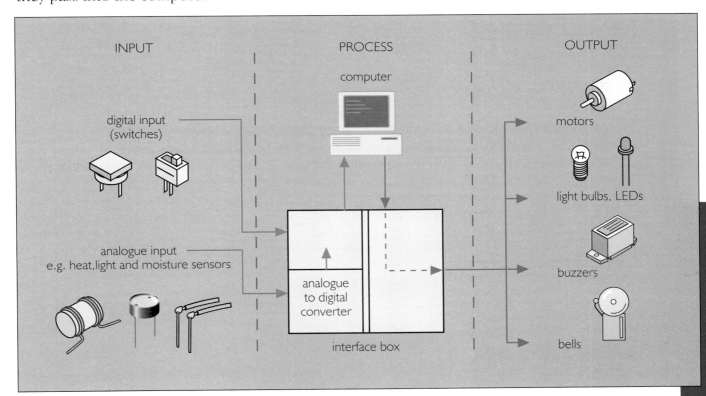

Programs

A computer uses **programs** to process the input – telling the computer what to do. A program needs to be written that includes all the operations to be performed.

Flow charts

The instructions given to the computer in the program need to be clear and follow a logical order. Drawing a flow chart makes writing a program easier.

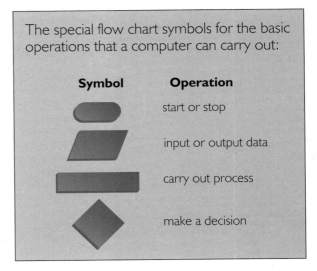

The special flow chart symbols for the basic operations that a computer can carry out:

Symbol	Operation
	start or stop
	input or output data
	carry out process
	make a decision

... Programs

A program can be a sequence of actions

This program is for an alarm. It flashes lights and sounds a buzzer.

You can make it into a more effective alarm by making the program repeat itself.

How could you turn off the alarm?

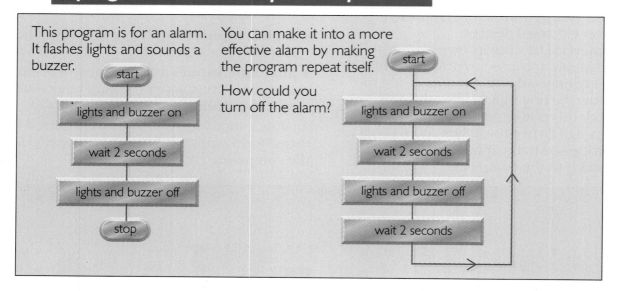

A program can respond to information from sensors

When you design a system to control movement, it is a good idea to include sensors that provide the feedback needed to stop the movement in the right place.

In this alarm program, the alarm repeats itself until a switch is pressed. This program could form a **procedure** within another program.

This program calls up an alarm procedure when an intruder sensor, like a reed switch on a window, is activated. What happens when the alarm is switched off?

This program, for a motor-driven sliding door, stops the motor as soon as the microswitch is pressed.

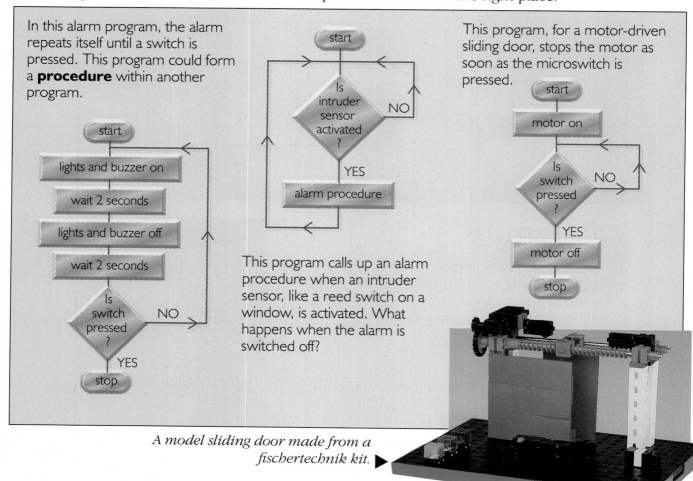

A model sliding door made from a fischertechnik kit.

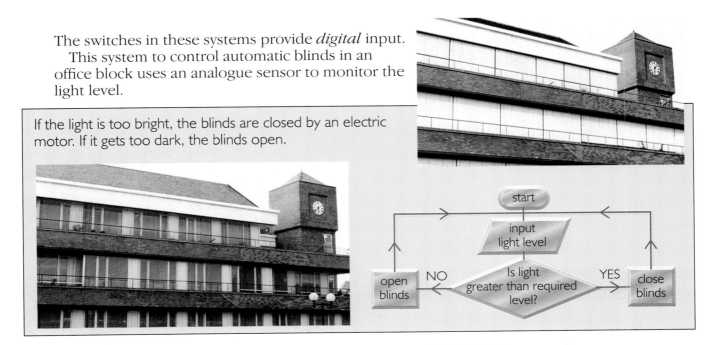

The switches in these systems provide *digital* input.

This system to control automatic blinds in an office block uses an analogue sensor to monitor the light level.

If the light is too bright, the blinds are closed by an electric motor. If it gets too dark, the blinds open.

A program can include a number of procedures

When you design a computer control system, it is helpful to break it down into subsystems and write a separate procedure for each one.

The main program calls up each procedure as needed. This program for a drinks machine shows how this works.

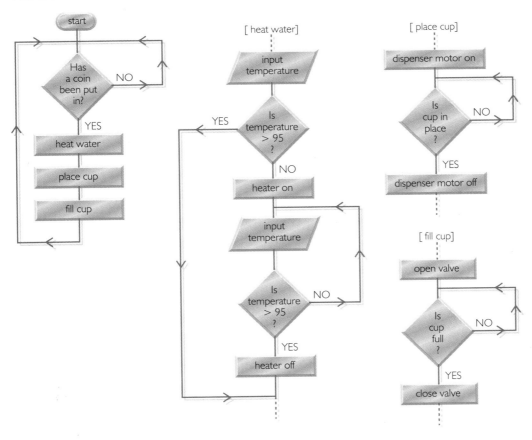

Notice how each procedure responds to feedback from an analogue or digital sensor.

Resource tasks

CCRT 1 – 3

8 COMPUTER CONTROL

Understanding structures

The arch bears up

A load such as a heavy tractor pushes downwards on this bridge, but the bridge stays up. This is how it works.

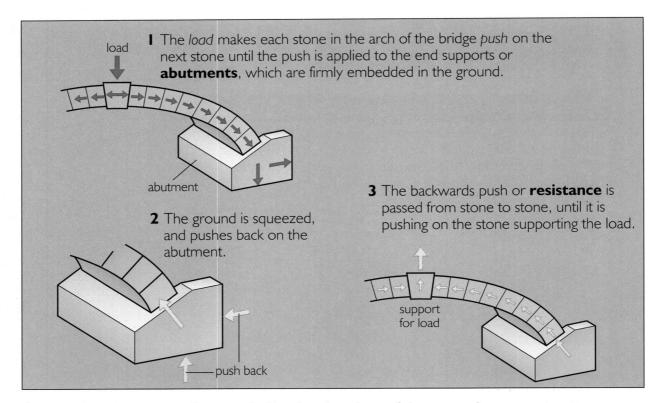

1 The *load* makes each stone in the arch of the bridge *push* on the next stone until the push is applied to the end supports or **abutments**, which are firmly embedded in the ground.

2 The ground is squeezed, and pushes back on the abutment.

3 The backwards push or **resistance** is passed from stone to stone, until it is pushing on the stone supporting the load.

If any stones are squeezed so much that they break, or if the ground is pushed away by the abutments, the bridge will collapse and the tractor fall into the river.

Forces

In these pictures forces (loads, pushes and pulls) are shown by red and yellow arrows.

Those passed on through a structure from a *load* are shown as red arrows.

Those showing *resistance* to loads because parts of the structure are being squeezed or stretched are shown as yellow arrows.

All parts of the bridge have forces on them coming both from the load and the resistance.

Engineers must foresee all these forces, and design and build the structure to ensure that each part is strong enough.

Resource task
StRT 1

Aloft on the aerial ropeway

The gang's hideaway is in the branches of an old beech tree. Now they are building a rope-and-pulley system for exciting rides. William has offered to test the rope.

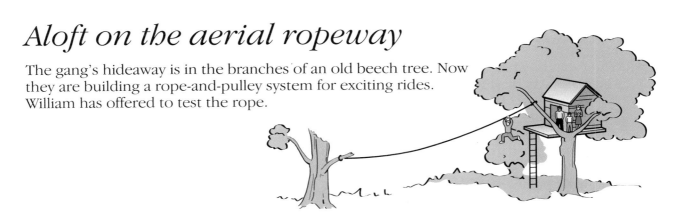

1 As soon as William hangs his full weight from the rope, it stretches. He drops a few feet and the rope becomes tighter.

2 The increase in tightness is passed on to the tree branches, making them bend. The bent branches pull backwards on the ends of the rope.

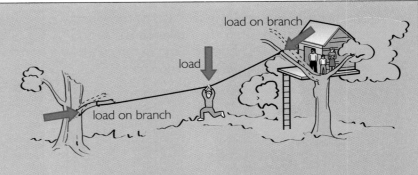

3 These pulling forces are passed back along the rope, right back to where William is hanging.

Ropes in tension

In a tug-of-war, the rope between the teams is stretched by a pulling force from each team.

Stretched, the rope tries to pull its ends inwards.

What happens if team A pulls harder than team B?

What happens if the rope is not strong enough?

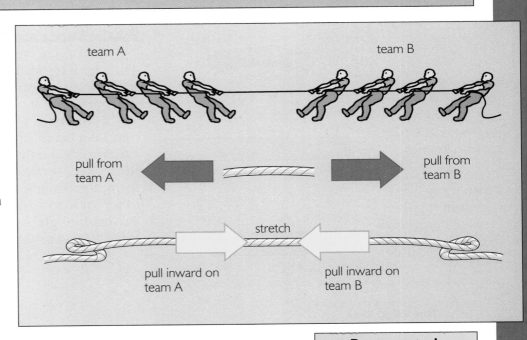

Resource task

StRT 2

Walking the plank

The cabin boy had been sentenced to 'walk the plank'. As he steps onto it, the plank begins to bend. The further he walks the more it bends, until ...?

The *load* of the cabin boy makes the plank bend.

The top of the plank is stretched and resists by pulling itself inwards. The bottom is squeezed and resists by pushing itself outwards.

The *pulling* effect along the top of the plank may be so strong that the fibres of the wood are torn apart. The plank snaps in two and the cabin boy is tipped into the sea.

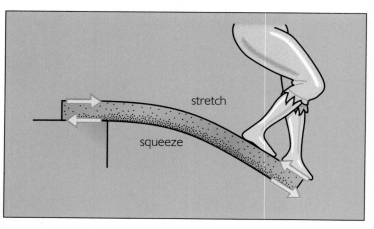

When a plank bends, one side stretches while the other side gets shorter.

When anything is stretched, it becomes tight (like the aerial ropeway), and pulls back on its ends.

When something is squeezed, it pushes outwards on its ends (like the stones in the arch bridge).

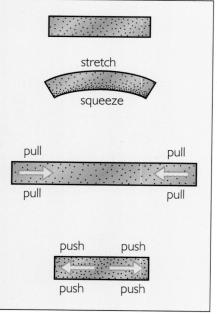

What happens to planks that bend?

The plank resists bending by pulling in on the ends along its top and pushing out along its bottom.

This happens whenever something bends. It may be a beam simply supported at each end or a cantilever (like the gang-plank). It may be horizontal, vertical or at an angle. Whatever the form, the designer has to foresee the effect of different load sizes and make sure the beam is strong enough to provide the pulling and pushing forces along its length without breaking.

Resource task
StRT 6

Upsetting the apple-cart

A market trader stacked his cart with apples. The apples sold fast and soon the barrow was nearly half empty.

Suddenly, the barrow tipped up and the apples poured onto the ground. Why?

Structures suddenly overbalance when the distribution of a load changes little by little.

The combined load of the weight of the apples and the cart causes upward resisting forces on the wheels and legs, from the ground.

As the apples are sold from the leg end of the cart, the combined load on the barrow moves away from that end until the upward resisting force is only on the wheels (and no longer on the legs). The cart is now exactly balanced.

Stable or unstable?

Structures usually need to be stable. The drawings below show how pairs of forces may cause things to topple over or to stay balanced.

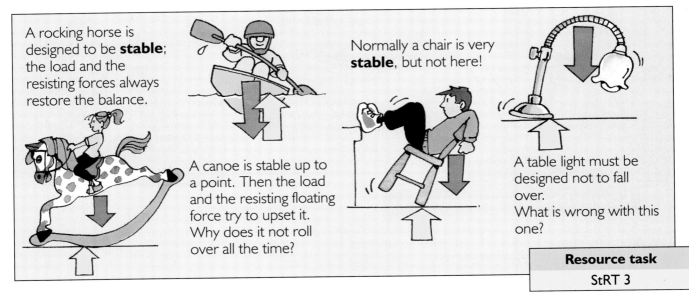

A rocking horse is designed to be **stable**; the load and the resisting forces always restore the balance.

A canoe is stable up to a point. Then the load and the resisting floating force try to upset it. Why does it not roll over all the time?

Normally a chair is very **stable**, but not here!

A table light must be designed not to fall over.
What is wrong with this one?

Resource task

StRT 3

Step-ladders that won't stay up

This painter tried to use an old step-ladder to reach the top of the wall. But whenever he tried to climb up, it slipped on the floor and went flat.

What was going wrong? As he climbed onto the steps, his weight provided a load, which was passed down, through the pieces of wood and their joints, to the floor. The floor, squeezed like the abutments of the arch bridge on page 178, responded with a resisting force upwards on the feet of the ladder, but....

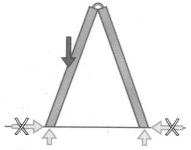
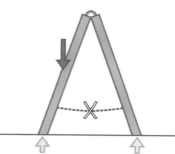

- there was no resisting force to stop the feet of the ladder from slipping out sideways;
- there was no connection between the parts of the ladder to stop it splitting apart;
- there was no stop on the hinge at the top to prevent its opening wide.

On a slippery floor any one of these would have been able to stop the step-ladder falling flat.

Then the painter tried to use the ladder for outdoor work. At first he succeeded. He set up the steps on the grass.

The feet of the ladder stuck into the soil, squeezed it and obtained a resisting force which stopped it moving any further.

Then he set it up half on grass and half on a path. This time there was resistance for one part of the ladder but not for the other... so, whoops!

Strengthening frameworks

Any framework, like this step-ladder, must have some way to keep its shape under a load. Almost always, there will be many possible ways of creating the necessary resisting forces.

Remember, the designer must always plan for at least one method of producing the necessary resisting forces.

Strength in hollow boxes

A well-made box has great strength. Box construction is often used in the design of heavily loaded structures such as ships, motor-car bodies and bridges. But it sometimes goes wrong, as these stories show.

The bus shelter that blew down
The wind blew off the back of the shelter. Then the walls and roof could no longer support each other and the shelter collapsed.

The car that crumpled
This box was designed to crumple. The energy used in crumpling the car during the crash is not available to injure the travellers.

The soap box that collapsed
Because one side of the box was missing the other sides began to sag. They could no longer support the load of the speaker.

The tower that toppled
The walls and floors of this tower block supported each other. When one wall was blown out, the rest collapsed.

What happens to loaded boxes?

The sides of boxes are very thin compared with their length and breadth. So why is a well-designed box so strong?

The sides are strong in tension. It is difficult to break a side by pulling along it. But it is easy to cause buckling by pushing on it.

Good design takes advantage of the strength in tension and ensures that the walls are supported to prevent buckling.

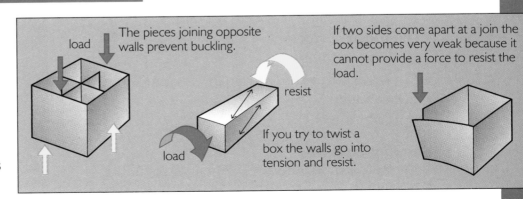

10 Designing and making with resistant materials

Choosing resistant materials

What are they like?

'Resistant material' is the term used for materials such as wood, plastic and metal. To describe a resistant material think about its physical properties, appearance and keeping qualities.

▲ *Can you describe and name the materials shown here?*

Physical properties
These govern what the material can do.
- **Strength** – a strong material will carry a heavy load without breaking.
- **Stiffness** – a stiff material will not bend or stretch easily.
- **Electrical conductivity** – an electrical **conductor** allows an electric current to flow through it.
- **Thermal conductivity** – a good **thermal** conductor allows heat to pass through it.
- **Heaviness** – a **dense** material will weigh a lot for its size.
- **Toughness** – a tough material withstands blows without breaking; a brittle material is easily broken.

▲ *The effect of surroundings on materials.* ▼

Appearance
You can describe the appearance of a material by asking questions:
- Does it look natural or manufactured?
- Is it rough or smooth?
- Is it shiny or dull?
- Is it **transparent** (see-through) or **opaque**?
- Is it coloured?
- Is it patterned?

Keeping qualities
Many materials are affected by their surroundings. Many metals are **corroded** by air and water. Wood is attacked by insects and fungus. Many plastics become discoloured and brittle with age. Some materials are meant not to keep, such as **biodegradable** materials.

Questions to ask yourself

When you are deciding which materials to use for your design you will need to consider all these questions. The information on pages 186–9 will help you. This example will help.

Tom chooses the materials for a steady-hand game

Tom had designed this game for his school fête. He had to choose the materials for each of the parts. He used the resistant materials information on pages 186–9. His choices and the reasons for them are shown below. Do you agree with him?

What does it need to do? What physical properties will be important?

What's available?

What materials have these properties?

How do I want it to look?

What keeping qualities will it need?

What's the best starting form?

How much can I afford?

10 DESIGNING AND MAKING WITH RESISTANT MATERIALS

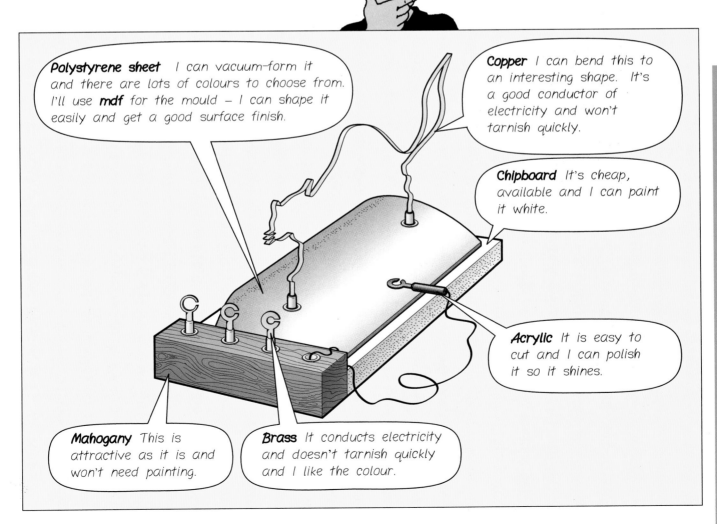

Polystyrene sheet I can vacuum-form it and there are lots of colours to choose from. I'll use **mdf** for the mould – I can shape it easily and get a good surface finish.

Copper I can bend this to an interesting shape. It's a good conductor of electricity and won't tarnish quickly.

Chipboard It's cheap, available and I can paint it white.

Acrylic It is easy to cut and I can polish it so it shines.

Mahogany This is attractive as it is and won't need painting.

Brass It conducts electricity and doesn't tarnish quickly and I like the colour.

...Choosing resistant materials

Consequences

In choosing materials, consider these important questions:

- Where does the material come from?

Some materials are grown and are renewable. Others are found underground and are not. The way both sorts of materials are obtained can harm the **environment**.

- Where will it go when the useful life of the product is over?

The design of your products should encourage the non-renewable materials to be recycled and reused. You can find out more about this in *Meeting needs and wants*.

The harmful effects of material production. ▶

Resistant materials information

Choosing wood

The information in the Chooser Charts and the available forms panel will help you decide whether to use wood and if so what sort.

Solid Timber Chooser Chart

Material	Important properties	Making tips	Cost	Typical uses
Red deal (often called pine)	softwood, cream and pale brown colour, often knotty, rots unless protected	moderately easy to cut, trim, shape and join	low	simple frameworks, block models
Jelutong	hardwood, light colour, no knots, more durable than red deal	easy to cut, trim, shape and join	medium	simple frameworks, block models, moulds for vacuum forming
Balsa	hardwood, whitish pink, very soft, very light, not durable	very easy to shape, cut and trim for joining use balsa cement	high	rapid model-making light-weight structures
Mahogany	hardwood, red-brown colour, durable	more difficult to work than red deal or jelutong	medium	containers, indoor furniture, items requiring decorative finish

Resource task

RMRT 1

Available forms

These forms of timber are readily available.

Manufactured Board Chooser Chart

Material	Important properties	Making tips	Cost	Typical uses
Plywood	tough doesn't warp exterior plywood is water-resistant	can split when cut	high	containers flat cut-out figures mechanical parts – links, cams, wheels
Hardboard	brittle goes soggy with water	tears easily difficult to finish edges	low	covering panels
Medium density fibreboard (mdf)	hard keeps edges well goes soggy with water	blunts tools shapes easily finishes well drills well	medium	block models vacuum forming moulds small bases
Chipboard	brittle edges easily damaged	difficult to shape blunts tools finishes poorly catches on drills	low	large bases

... Resistant materials information

Choosing metals

The information in the Metals Chooser Chart and the available forms panel will help you decide whether to use metal and if so what sort.

Metals Chooser Chart

Material	Important properties	Making tips	Cost	Typical uses
Mild steel	silver-grey colour, stiff and strong, rusts in moist air, ferrous, alloy of iron and carbon	easy to join using heat (brazing), difficult to deform or melt and cast, quite hard to shape	low	mechanical parts such as axles and linkages, frameworks from both strip or tube
Aluminium	silver-white colour, low density, non ferrous	difficult to join using heat, easy to deform, shape and cast	medium	castings for jewellery, decorative items and fittings
Copper	pinkish-brown colour, good conductor, tarnishes slowly in moist air, non ferrous	easy to join using heat (solder), very easy to deform and shape	high	decorative items, electrical contacts
Brass	yellow colour, hard, tarnishes slowly in moist air, alloy of copper and zinc, non ferrous	easy to join using heat (solder), fairly easy to cast	high	mechanical parts such as couplings and bearings, decorative items

Available forms
These forms of metal are readily available.

Choosing plastics

The information in the Plastics Chooser Chart and the available forms panel will help you decide whether to use plastics and if so what sort.

Plastics Chooser Chart

Material	Important properties	Making tips	Cost	Typical uses
Acrylic	stiff and strong but not tough scratches easily wide range of colours **thermoplastic**	good for strip heating polishes well join using Tensol cement	medium	containers and storage devices flat cut-out figures mechanical parts – links, cams, wheels
PVC (poly vinyl chloride)	stiff, strong and tough more scratch-resistant than acrylic **thermoplastic**	join using liquid solvent cement (sold as plumbers' material)	medium	containers and storage devices
Polystyrene (high impact polystyrene)	not tough wide range of colours **thermoplastic**	good for vacuum forming join using liquid polystyrene cement	low	shell forms for containers, model boats, model cars
ABS (acrylonitrile butadienestyrene)	stiff, strong and tough scratches easily wide range of colours **thermoplastic**	easy to cut and trim join using liquid solvent cement	medium	frameworks and mechanical parts – links, cams, wheels
Nylon	stiff, strong and tough self-lubricating **thermoplastic**	machines well difficult to join with adhesives	high	good for bearings and mechanical components
Polyester resin	liquid, sets to a hard solid Wide range of colours **thermosetting** plastic	important to use the correct amount of catalyst for hardening	medium	solid, decorative castings reinforced with glass fibre to give strong shell structures

Available forms
These forms of plastics are readily available.

Work on it safely

Think about all of these things to make sure that you are safe when you make your design. Observe these four important rules:

1 All students should wear a protective apron or overall for D & T.
2 When using machines, students must use eye-protectors.
3 Only one student should use a machine at any time.
4 If in doubt ask the teacher.

Here are some instructions for personal readiness and safety in your workplace.

Personal readiness
- Tuck in loose clothing.
- Tie back long hair.
- Remove jewellery.
- Use safety goggles, gloves, face mask and ear protection as instructed.
- Move around carefully, especially when carrying things.

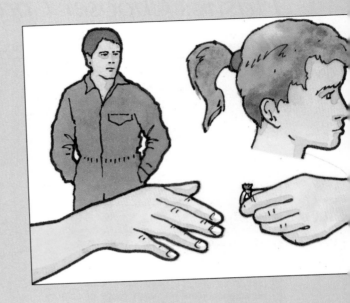

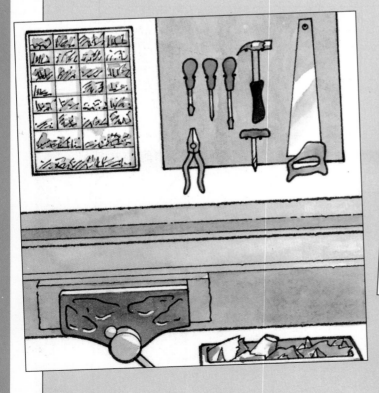

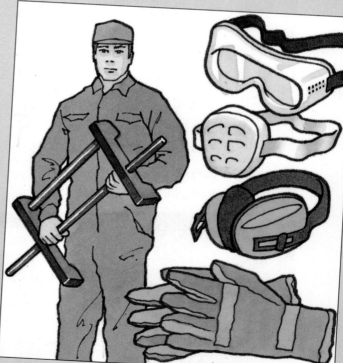

The place where you work
- Keep your area tidy.
- Clean away waste materials.
- Use the correct work areas for particular processes, e.g. heating metal.

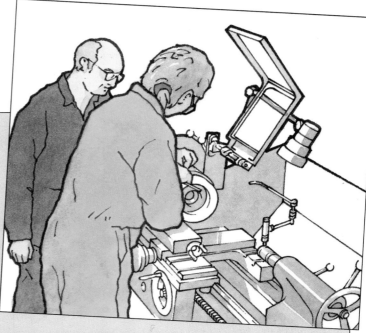

Materials you intend to use
- Some are heavy if dropped: take care.
- Wood has splinters: take care.
- Metal and plastics can have sharp edges and corners: handle carefully.
- When sanding wood or plastics wear goggles and a face mask, and make sure that the extraction system is on.
- Hot materials can cause burns: wear gloves.
- Sticky materials can make a mess: use a brush or spreader.
- Fumes may be produced: check ventilation.

Machine tools
- Make sure your work is held securely in the machine.
- Don't try to carry out a process if you have not been shown how to do it.
- Do not queue to use a machine.
- Follow instructions carefully and ask if you are not sure.
- Do not adjust a machine if it is still moving.
- Be sure to remove chuck keys before switching on.
- Check guards are in place before switching on.
- Do not distract anyone using a machine.

Hand tools
- Check that you are using *exactly* the right tool for the job.
- Check that edge tools are properly sharp.
- Always keep both hands away from cutting edges.
- Keep fingers clear of danger areas when using hammers, mallets or screw-drivers.
- Do not distract anyone using hand tools.

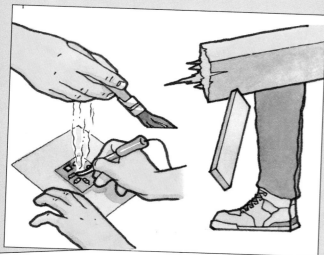

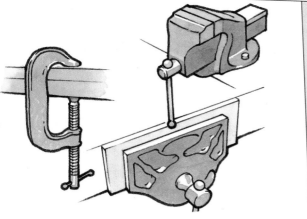

Holding
Use a vice, cramp or jig to secure the material so that both hands are free to work safely.

Marking out and checking that it's right

You will need to transfer your designs to the materials carefully so that you can cut and shape accurately. Unless you mark out and check thoroughly you will not be able to make your design well.

Check for the straightest part first

Usually your material will have at least one surface or edge which is already accurate. Check for this with a steel rule. Mark out using this surface or edge as a starting point.

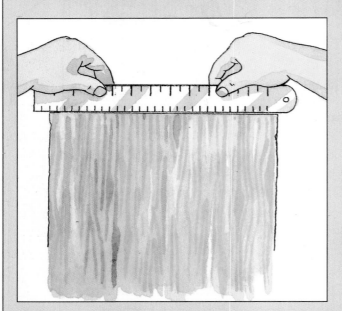

Marking and checking angles

Tools to help you mark out and check 90° angles:

- For wood use a try-square.
- For metal or plastics use an engineer's square.
- Draw or check angles other than 90° with an adjustable bevel.

Making marks

Use different markers for different materials. Use a sharp pencil for wood or the paper protection on plastic sheet.
Use a scriber or spirit-based felt pen for metal or bare plastics.

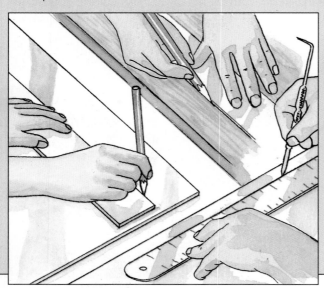

Marking out lines parallel to a straight edge

Use a marking gauge on wood.
Use odd leg callipers on metal or plastic sheet.

Marking out and checking circular parts

Mark circular curves with a pencil compass on wood or paper protection on plastic.
For metal use a pair of spring dividers and mark the centre with a centre punch.

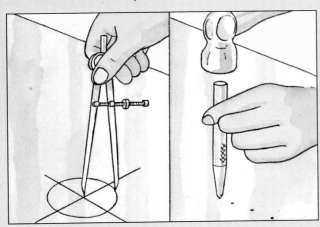

Check the inside and outside diameters of circular objects with inside or outside callipers which can then be measured against a rule.

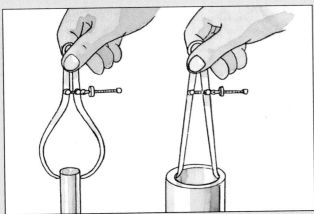

Use a centre square to draw in a diameter. To find the centre of an 'unknown' circle, draw two diameters and see where they cross.

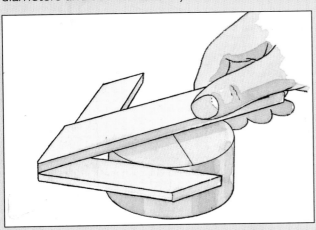

Marking out irregular shapes

To mark out irregular shapes use a card template. They are useful if you want to mark out more than one shape.

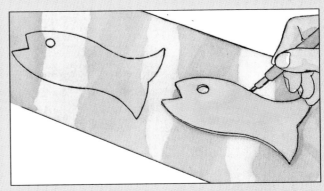

Marking out the position for drilling holes

To drill a hole you only need to mark out the centre. Use a small, accurate cross.
To stop the drill bit from wandering away from the centre of the cross, mark it with a centre punch on metal and a bradawl on wood.
For plastics mark the cross on a piece of masking tape.

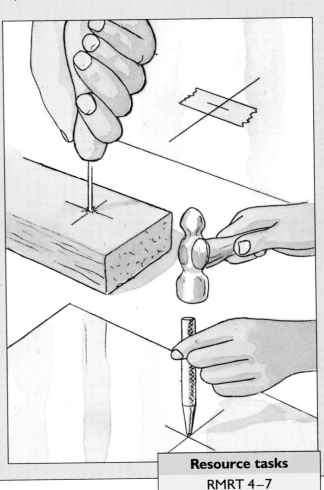

Resource tasks
RMRT 4–7

10 DESIGNING AND MAKING WITH RESISTANT MATERIALS

193

Cutting the pieces and trimming them

The cutting tool you choose to use will depend on the material you are cutting – wood, metal or plastic – and whether you want to cut a straight line or a curve.

Sawing

Hard or thin materials have to be cut with a saw which has many small teeth. Thicker and softer materials can be cut with a saw with larger teeth.

Sawing straight

Use a tenon saw for wood and a hacksaw for metal and plastic. The blades are stiff and wide and cannot be twisted to follow curves.

Sawing curves

Use a coping saw to cut curves in wood and plastics. The blade is set so that it cuts when it is pulled.

Use an abra file for metal. The blades are narrow so that you can turn them around the curve.

You can also use a small machine fret saw. This will cut thin wood and metal. To cut plastics, cover the part to be cut with masking tape to stop the plastic's waste sticking in the cut behind the blade.

▲ A tenon saw cuts a piece of wood held against a bench hook.

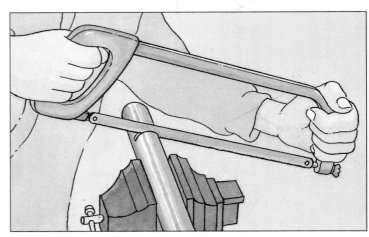
▲ A hacksaw cuts a mild steel bar held in an engineer's bench vice.

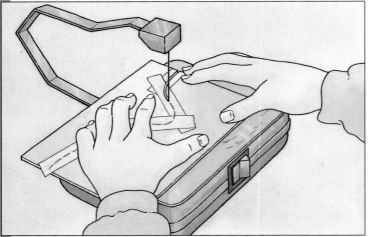
▲ A small machine fret saw cuts a curve.

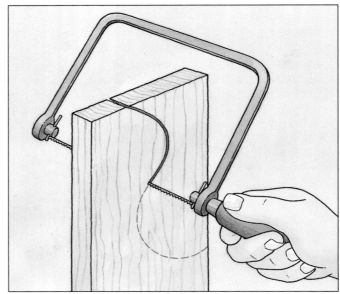
▲ A coping saw cuts a curve.

Shearing

Thin sheet metal can be difficult to cut with a saw. You can cut it more easily by using tin snips which **shear** through the metal.

Trimming

It is difficult to saw or shear materials accurately. To get the exact shape you want you will need to **trim** the material down to the line after it has been cut off or cut out.

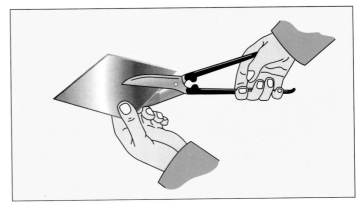

Using chisels

Use a chisel to clear away material neatly and make slots and grooves.

Remember the following safety points:

- Hold the work firmly in a cramp or vice, leaving both your hands free.
- Use a wooden mallet to tap the chisel, so that you don't make your hand sore.
- Keep both hands behind the cutting edge of the chisel. Then, if it slips, your fingers are safe.

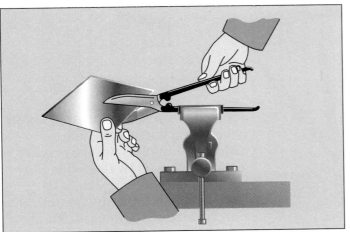

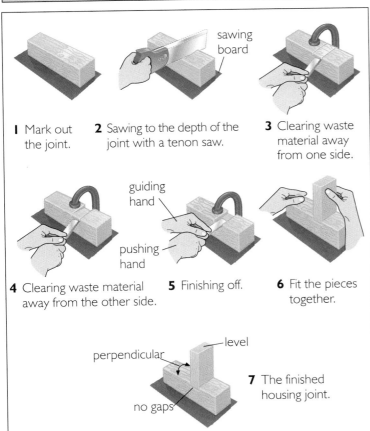

▲ *Making a simple housing joint.*

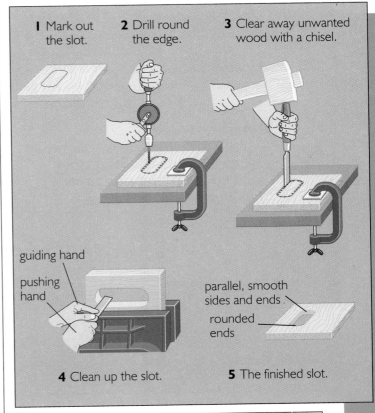

▲ *Cutting a slot.*

Resource tasks
RMRT 3, 4, 5, 8

...Cutting the pieces and trimming them

Trimming metals and plastics

Trim metals and plastics with a file. They are available in a range of size of teeth, length and cross-sectional shapes.

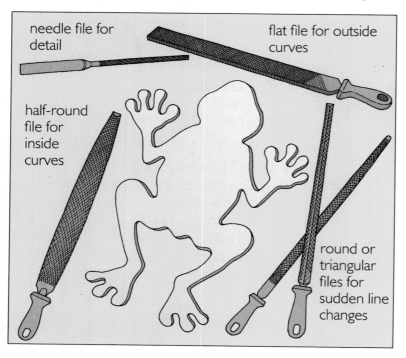

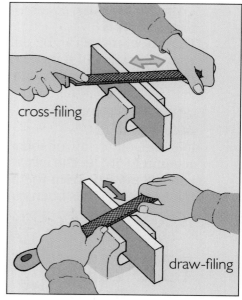

Which file you use will depend on the shape you are making. Most trimming is done by cross-filing, followed by draw-filing to a good finish.

Sawing and shearing materials is an approximate process. For an accurate shape, trim the material down to the line after it has been cut off or cut out.

▶ *Trim the straight edges on a piece of wood or manufactured board with a plane.*

Trimming wood

To plane the end grain of solid wood you must plane in half-way from each end to prevent the corners from breaking.

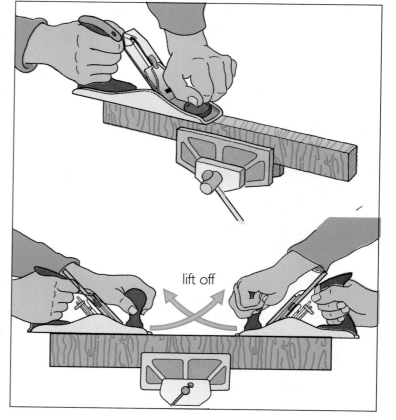

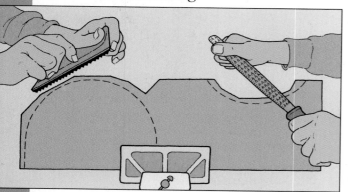

▲ *For inside and outside curves use a surform or wood rasp.*

Resource tasks
RMRT 3, 4, 5, 8

Using a sanding machine

Small pieces of wood and plastics can be trimmed on a sanding machine. You can make a simple **jig** to trim circular pieces of material. Remember to mark out a line to which you can work.

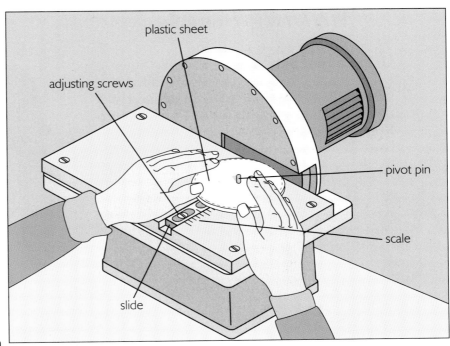

A sanding machine being used to trim a circle. ▶

Making holes

Drilling holes

Using hand drills

You can drill small diameter holes (1–6 mm) by hand with a wheel brace (hand drill) and twist drill. This method is suitable for small holes in wood, plastics and softer metals.

Hold the work firmly so it does not move or bend.

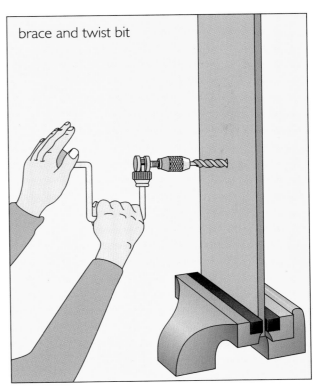

Put scrap wood under your workpiece to prevent the material splitting or tearing.

Larger holes in wood can be made with a brace and twist bit. In this case bore until the point just breaks through.

Turn the work over and, using the small hole made by the point as a guide, bore through from the other side.

... Making holes

Using a machine drill

Using a machine drill, sometimes called a pillar drill, makes drilling much easier. Hold the material in a vice or **jig** or with a **cramp**. Rest it on scrap wood to prevent splitting or tearing.

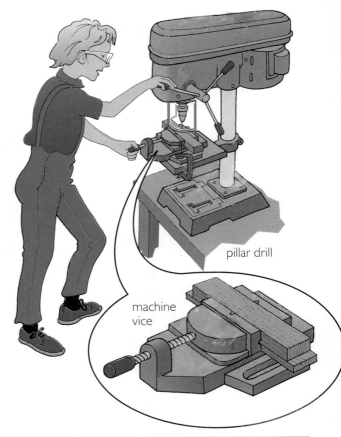

pillar drill

machine vice

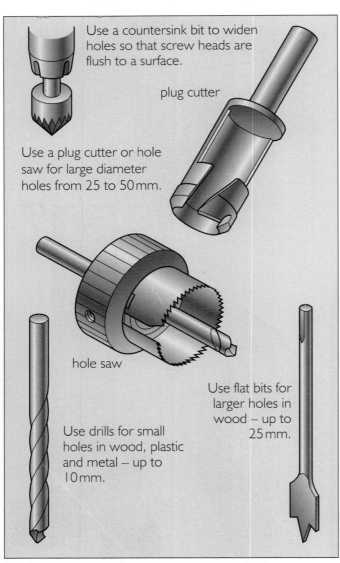

Use a countersink bit to widen holes so that screw heads are flush to a surface.

plug cutter

Use a plug cutter or hole saw for large diameter holes from 25 to 50mm.

hole saw

Use drills for small holes in wood, plastic and metal – up to 10mm.

Use flat bits for larger holes in wood – up to 25mm.

Sawing holes

You can use a coping saw or abra file. These can be taken apart and reassembled with the blade through a hole so that a shape can be cut from the centre of the material.

Resource tasks
RMRT 3, 4, 5, 8

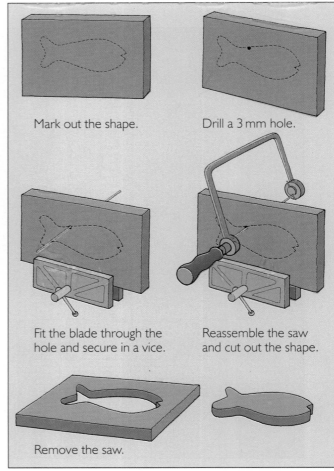

Mark out the shape.

Drill a 3 mm hole.

Fit the blade through the hole and secure in a vice.

Reassemble the saw and cut out the shape.

Remove the saw.

Forming with resistant materials

Folding and bending

You can use simple folds or bends in some materials to make interesting forms.

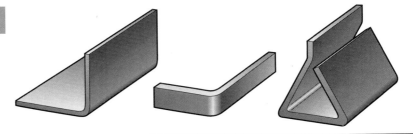

Bending thermoplastics with a strip heater

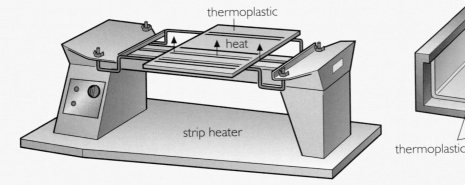

Mark the position of the fold onto the outside of the plastic with a marker or chinagraph pencil.

Heat both sides of the material where it is marked, using a strip heater.

When it feels soft enough, fold it to the required angle. If it is hot, use protective leather gloves.

When cool the plastic will retain its new shape and the marking out can be cleaned off.

For accuracy check the angle and use a jig.

Bending sheet metal with a vice and folding bars

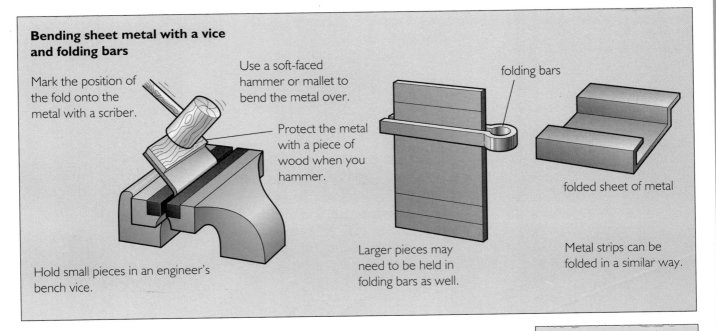

Mark the position of the fold onto the metal with a scriber.

Use a soft-faced hammer or mallet to bend the metal over.

Protect the metal with a piece of wood when you hammer.

Hold small pieces in an engineer's bench vice.

Larger pieces may need to be held in folding bars as well.

Metal strips can be folded in a similar way.

Resource task

RMRT 3

...Forming with resistant materials

Forming sheet materials

You can form some sheet materials into shell forms by vacuum forming, hollowing and plug-and-yoke forming.

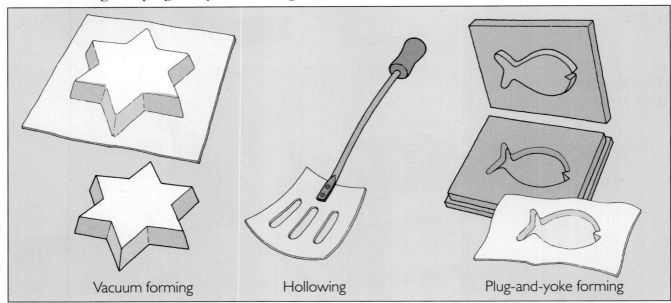

Vacuum forming   Hollowing   Plug-and-yoke forming

Plug-and-yoke forming thin thermoplastics sheet

1 Mark out the required piece of mdf or plywood.

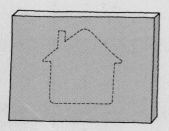

2 Cut around the shape with a coping saw starting in from the outside. Make sure that the edges are made smooth. The piece you cut out is the plug, the piece you leave becomes the yoke.

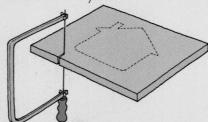

3 Glue another piece of plywood or **mdf** underneath the yoke.

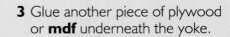

4 Warm up a piece of thermoplastics sheet in an oven. When it is soft and flexible lay it over the yoke and press the plug into the yoke.
Wear protective leather gloves.

5 When it is cool the plastic will retain its new shape. Cut away the excess material, and trim the edges. Then smooth them using a sheet of glasspaper on a flat work-top.

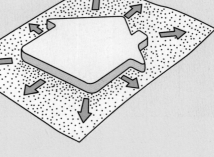

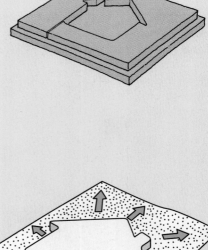

Vacuum forming thermoplastics sheet

1 Prepare a mould of the form you want to make. Use plywood, solid wood or mdf.
  The sides of the mould must slope slightly to help remove the form from the mould after completion. The mould should be smooth and polished.
  Moulds can also be made from dried and hardened clay.

2 Place the mould on the bed of the vacuum-forming machine.

3 Clamp the sheet of thermoplastics into place. Heat it until it is soft and flexible.

4 The machine will bring the mould and the thermoplastics sheet together. With the vacuum pump withdraw the air from between the mould and thermoplastics sheet. Air pressure on the outside will press the sheet against the mould.

5 When it cools, the thermoplastics sheet will keep its new shape. Tap out the mould, cut away the excess material and trim and smooth the edges.

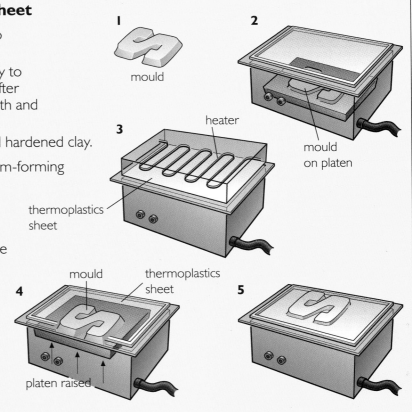

Forming sheet aluminium alloy or copper by hollowing

1 Mark out and cut the required outline shape from the flat sheet with tin snips.

2 **Anneal** the metal by heating it. This makes it easy to work. Copper will glow cherry red when it is hot enough. Aluminium alloy needs less heat but gives no tell-tale glow. Smear some soap onto the cold metal and heat it until the soap goes black.

3 Cool the metal in cold water. Form a dish by hammering the metal into a sandbag with a bossing mallet. Start from the outside and work into the middle.

4 This will give a rough shape which can be made true by using a blocking hammer to hammer the form into a hollow carved into the surface of a heavy piece of wood. Or hammer the shape from the outside with a flat-faced mallet over a domed metal stake.

5 Clean up the metal and polish it.

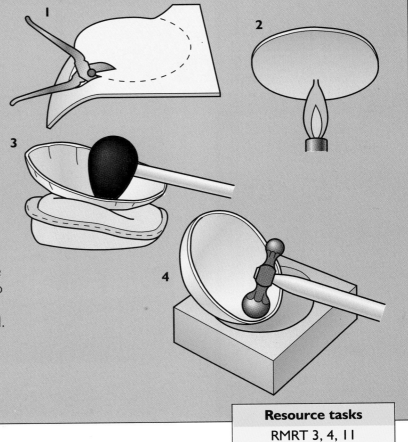

Resource tasks
RMRT 3, 4, 11

Casting

Here are some examples of products that have been **cast**. In each case a material in liquid form has been poured into a **mould** and set solid to form a **casting**.

▲ *All these objects have been produced by casting.*

Casting with metal

1. Melt the metal by heating it in a ladle.
2. Pour the molten metal into the mould. It takes up the shape of the mould.
3. The metal cools in the mould and freezes solid.
4. Split open the mould to get the casting out.
5. Trim the spare metal off the casting.
6. You can paint the finished casting.

Casting with resin

1. Mix the liquid resin with the correct amount of catalyst and any colouring.
2. Pour the mixture into the mould.
3. The liquid resin sets solid to form a thermosetting plastic.
4. Strip the mould off the casting.

Resource task

RMRT 7

Using machine tools

You can use lathes and milling machines to make components with complex shapes.

The lathe

Using computer-controlled tools

You may be able to use a machine tool that is controlled by micro-computer. You can produce the design on-screen using a computer. This is CAD – computer assisted design.

The computer uses this information to control the cutting machine that makes the design. This is CAM – computer aided manufacture.

The mill

10 DESIGNING AND MAKING WITH RESISTANT MATERIALS

Joining

Describing joining

Four important terms you will need to use when describing how two pieces of material are joined together are:
- **temporary**;
- **permanent**;
- **rigid**;
- **flexible**.

These examples will help you understand what they mean.

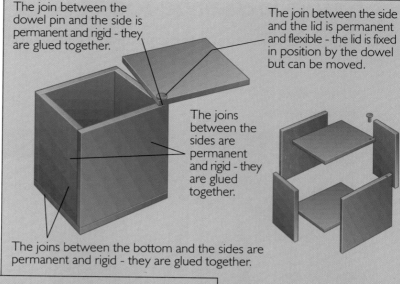

The join between the dowel pin and the side is permanent and rigid - they are glued together.

The join between the side and the lid is permanent and flexible - the lid is fixed in position by the dowel but can be moved.

The joins between the sides are permanent and rigid - they are glued together.

The joins between the bottom and the sides are permanent and rigid - they are glued together.

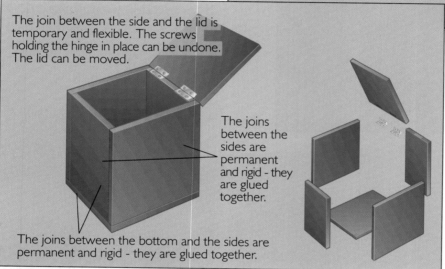

The join between the side and the lid is temporary and flexible. The screws holding the hinge in place can be undone. The lid can be moved.

The joins between the sides are permanent and rigid - they are glued together.

The joins between the bottom and the sides are permanent and rigid - they are glued together.

Adhesives

You will often need to join materials together using **adhesives**. The Adhesives Chooser Chart will help you make the right choice.

Adhesives Chooser Chart

Adhesive	Uses
PVA (polyvinyl acetate) e.g. Evostik Resin W	a general-purpose wood glue; not water-resistant
Synthetic resin e.g. Cascamite	for joining wood; waterproof and stronger than PVA; must be made up immediately before use
Epoxy resin e.g. Araldite	for joining metals and acrylic plastics; waterproof; must be made up immediately before use
Contact adhesive* e.g. Dunlop Thixafix	for joining polystyrene, fabrics and leather
Acrylic cement* e.g. Tensol	for joining acrylic plastics

*Must be used in a well-ventilated area.

Fittings

You will often need to fix materials together using **fittings** such as nails, screws or nuts and bolts. The Fittings Chooser Chart will help you make the right choice.

Fittings Chooser Chart

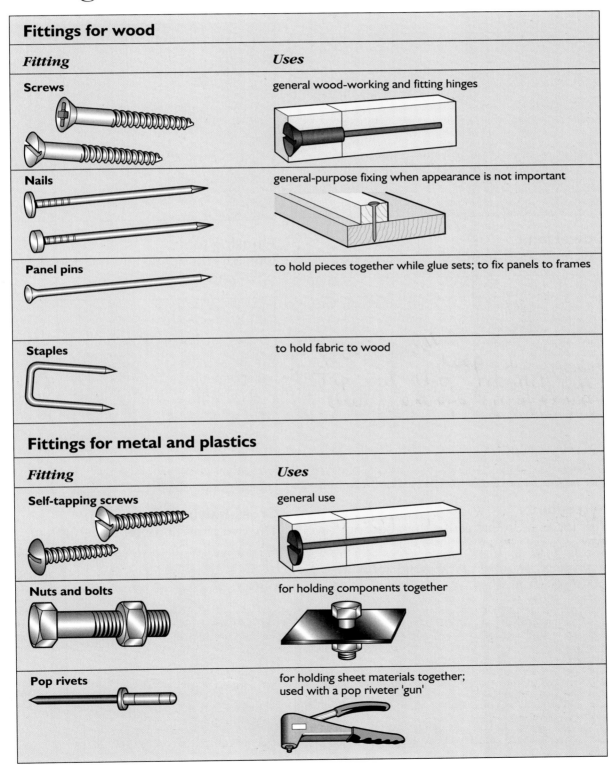

Fittings for wood	
Fitting	*Uses*
Screws	general wood-working and fitting hinges
Nails	general-purpose fixing when appearance is not important
Panel pins	to hold pieces together while glue sets; to fix panels to frames
Staples	to hold fabric to wood

Fittings for metal and plastics	
Fitting	*Uses*
Self-tapping screws	general use
Nuts and bolts	for holding components together
Pop rivets	for holding sheet materials together; used with a pop riveter 'gun'

...Joining

Ways of joining wood

Here are some different ways of joining pieces of timber or manufactured board.

Simple butt joint
The two pieces have been pinned and glued together.

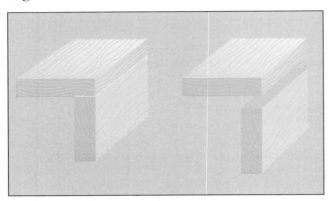

Lapped joint
This is stronger than a simple butt joint.

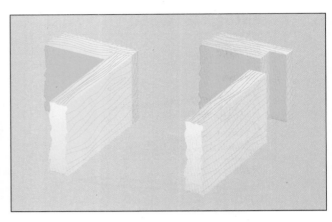

Dowel joint
This is stronger than a simple butt joint.

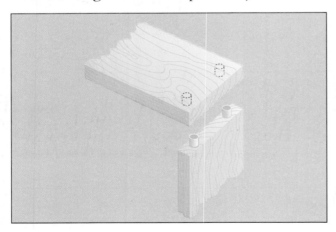

Housing joint
One part fits tightly into the housing in the other part.

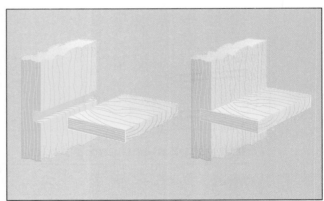

Mitre joint
This looks more attractive than butt or dowel joints.

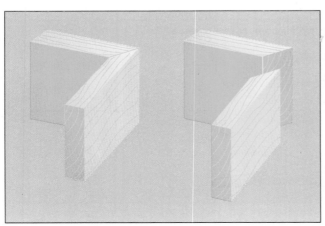

Cross-halving joint
The two parts fit together tightly.

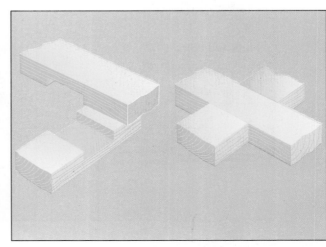

Brazing and soldering

You can join two metals by using a different metal with a lower melting point as the 'adhesive'. When you heat the joint, the low melting point metal melts and flows into the gap. As the joint cools this metal turns solid, joining the other two pieces of metal together in a permanent, rigid joint.

It is important to clean the pieces of metal to be joined. Use a **flux** to keep them clean when heating the joint. If they are dirty, the liquid metal will not stick to them properly so the joint will be weak or will not form at all.

The table shows the different solders you can use.

Materials to be joined	Solder	Strength of joint
Copper, brass or tin-plate	soft solder m.p. 185 – 230 °C	weak
Copper or steel	silver solder m.p. 720 – 800 °C	strong
Copper or steel	brazing spelter m.p. 900 °C	very strong

▲ Soft-soldering tin-plate.

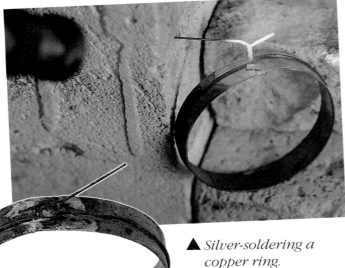

▲ Silver-soldering a copper ring.

▲ Brazing a steel cross-piece.

Assembling different parts

You can often make a product by assembling parts. You might use wood screws or nuts and bolts, so that it can be taken apart again. Here are three other methods of assembly.

Producing a carcass with knock-down fittings

1 Mark out, cut and trim the panels needed for the **carcass.**
2 Hold the carcass together using cramps.
3 Place the knock-down joints in position and mark out.
4 Take the carcass to pieces and attach the fittings.
5 Reassemble using the fittings.

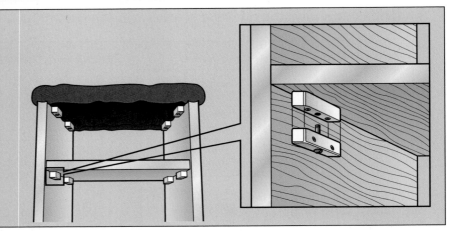

Producing frameworks

A framework from wood strip

1 Mark out, cut and trim the lengths of wooden strip needed for the frame.
2 Mark out and cut joining corners from stiff card or thin plywood.
3 Hold the framework together using cramps.
4 Apply glue to the joining corners and place in position on the framework. Leave to dry. When dry, turn it over and repeat the process.

You may have to work in stages if the framework is complicated.

A framework from PVC tubing

1 Mark out, cut and trim the lengths of tubing needed. Use a file to remove the sharp edges at the ends.
2 Assemble the framework with the corner and T fittings, without adhesive first to check that everything is correct.
3 Take it to pieces and reassemble using a plastic adhesive.

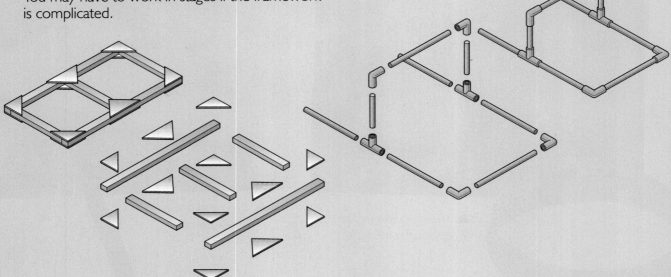

Making resistant materials look good and stay good

Heat treatment

Hardening and tempering

You can **harden** tool steel by heating it to red and then **quenching** (cooling) it in water or oil. This also makes the steel brittle. If parts of the steel need to be tough as well as hard then you need to re-heat it carefully in a process called **tempering**.

To temper a piece of steel heat it very gently so that the surface gets hot enough to combine with the oxygen in the air and form a layer of coloured oxide. Different colours of oxide are formed at different temperatures. The colour of the oxide formed tells you the temperature. The steel gains different properties at different temperatures.

Tempering is used to give steel tools the properties they need as shown in the panel.

▲ *Careful heating tempers steel.*

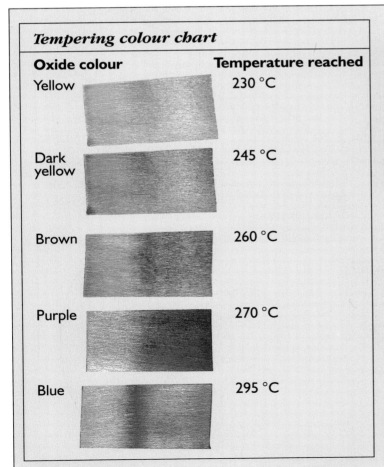

Tempering colour chart

Oxide colour	Temperature reached
Yellow	230 °C
Dark yellow	245 °C
Brown	260 °C
Purple	270 °C
Blue	295 °C

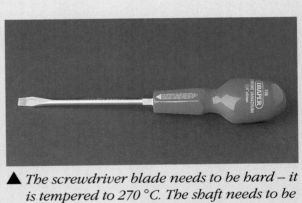

▲ *The screwdriver blade needs to be hard – it is tempered to 270 °C. The shaft needs to be tough – it is tempered to 230 °C.*

▲ *This steel tape needs to be springy – it is tempered to 295 °C.*

...Making resistant materials look good and stay good

Cleaning up

You apply a finish to a resistant material to protect it from the environment and to make it look good.

The first step in applying a finish, called cleaning up, is getting the surface of the material as smooth as possible.

Any surface or edge on which you have used tools will be marked. You should remove these marks.

You can use **abrasives** to wear down the material. Start with coarse abrasive to remove the deepest marks and work down to a fine abrasive, which should leave the material smooth and free of marks.

To clean wood
Use glass paper wrapped around a cork block. The finest grades of paper are called flour paper. Work with the direction of the grain. If you cut across the grain it will leave scratches that are difficult to remove.

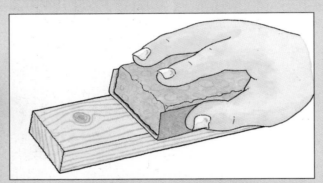

To clean metal
Use emery cloth, working down through the grades. It is often wrapped around a file to keep it flat. A drop of oil will help to remove the waste metal produced.
Final polishing of copper and aluminium alloy can be done with a buffing machine or by hand with metal polish.

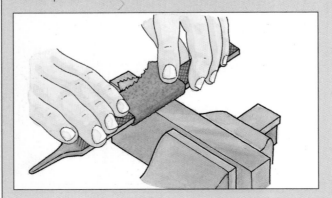

To clean plastics
Use 'wet and dry' (silicon carbide) paper, with water. The water helps remove the waste plastic and stops it clogging up the paper. It also prevents friction heating and spoiling the surface.

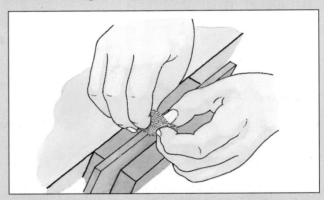

A tightly rolled ball of 0000 grade steel wool can be used as an alternative.
Final polishing can be done with a buffing machine or by hand with metal or plastics polish on a cloth or leather pad.

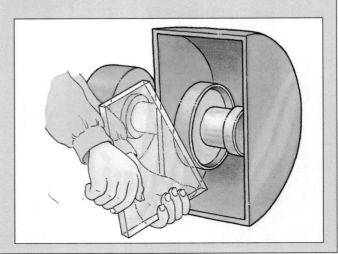

Applying a finish

Most woods and some metals will need to have a finish applied. There are two basic types:

- a substance that coats the surface forming a protective skin;
- one that is applied to the surface and soaks into the material.

Plastics do not require a finish other than polishing.

Protective skin finishes

Paints, varnishes and lacquers

These form protective skins on wood and metal. Paints usually hide the material and provide the surface decoration as well as protection. Some paints require an undercoat.

Varnish and lacquer are usually transparent so the material they are applied to can show through. You can apply them with a brush or a spray.

Leave them to dry somewhere undisturbed and dust-free so that dirt and finger marks do not spoil the finish.

Enamelling

Use enamelling for a highly coloured, tough and attractive finish on copper.

To enamel onto copper:

1 Mark out, cut and trim the required shape of copper blank.
2 Slightly dome the blank with a bossing mallet.
3 Clean the copper blank thoroughly with emery cloth.
4 Place the blank on paper. Mix the enamel powder to a paste with water. Paint the paste onto the blank.
5 Heat the blank in a kiln or on a mesh over a flame until the enamel powder fuses.
6 Set the piece aside to cool down slowly.

Resource tasks
RMRT 8, 9

...Making resistant materials look good and stay good

Dip-coating

Use dip-coating for a coloured finish that protects metals from air and moisture.

To dip-coat a piece of mild steel with plastic:

1 Ensure that there is a means of attaching a wire hanger to the object.

2 Clean the object with emery cloth.

3 Hang the object in an oven set at 180° Celsius.

4 Carefully dip the object into a fluidizing tank containing polythene or nylon powder. **A face mask is essential**.

5 Put the object back in the oven until the grains of plastic fuse together.

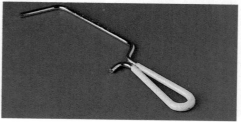

Finishes that soak in

Each of these products uses a finish that soaks in.

▲ The oil used here makes the wood darker and protects it from water.

▲ This stain decorates but does not protect the wood.

▲ The finish used here protects but does not alter the appearance.

This chart will help you choose the finish needed for your product.

Finishes Chooser Chart

Finish	Does it alter appearance?	Does it protect the material?	Can I use it on wood?	Can I use it on metal?
Paint	Yes	Yes	Yes	Yes
Varnish	No	Yes	Yes	No
Lacquer	No	Yes	Yes	Yes
Enamelling	Yes	Yes	No	Yes
Dip-coating	Yes	Yes	No	Yes
Coloured stain	Yes	No	Yes	No
Linseed oil	No	Yes	Yes	No
Sanding sealer	No	Yes	Yes	No
Oil quenching	Yes	Yes	No	Yes

Resource tasks
RMRT 8, 9

Choosing tools

The information in the Tools Chooser Chart will help you choose the right tools for the job.

Tools Chooser Chart

Process	Wood	Metal	Plastics
For marking out	pencil	scriber	felt-tip pen or scriber
• at right angles	try-square	engineer's square	engineer's square
• parallel to an edge	marking gauge	odd-leg callipers	odd-leg callipers
• an irregular shape	card template	card template	card template
For holding	woodwork vice	metalwork vice	metalwork vice
	G-cramp	G-cramp	G-cramp
	machine vice	machine vice	machine vice
For cutting			
• straight lines	tenon saw	hacksaw	hacksaw
• curves	coping saw	tin snips	abrafile
	fret saw	abrafile	coping saw
For trimming			
• to a straight line	plane	flat file	flat file
	sanding machine		sanding machine
• to a curve	rasp	flat file	flat file
	surform	round file	round file

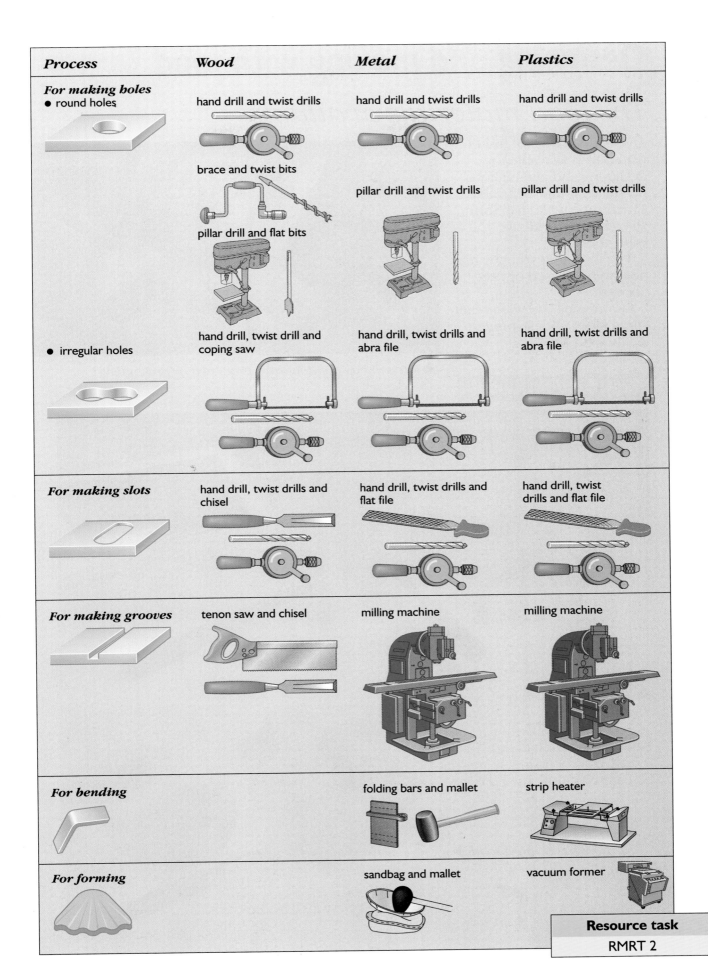

11 Designing and making with food

The food materials available

An exciting range of food materials is available for designing and making food products. Fruits, vegetables, cereals, meat and fish from all over the world give you a wide choice.

The food that you buy may be raw, processed or pre-cooked.

Sometimes you will use all three types in designing a single food product.

▲ You can buy potatoes raw, processed or pre-cooked.

Raw food materials

You can change the appearance of raw food by peeling, chopping or grating.

whole   peeled   chopped   grated

▲ Raw carrots come in several forms.

Raw foods have had little or no industrial processing. Some can be eaten without any cooking.

Cooking changes food. ▶

raw   cooked

You can change the colour, texture and taste of raw food by cooking it.

These processed foods can be eaten without cooking.

Processed foods

Processed foods are made by manufacturers from raw foods. Some can be eaten in the form you buy them. Others must be cooked.

Pre-cooked foods

These have already been cooked, like bread or biscuits. You can eat them as they are or use them to make a different food product.

The food industry produces some pre-cooked foods as **convenience foods**. These are complete dishes or meals that only need reheating.

These processed foods must be cooked before you eat them.

▲ *Pre-cooked and convenient!*

The properties and qualities of food

Food materials have properties and qualities that you can control when preparing and cooking. You will need to think about these when you are designing food products. This section will help you to describe the physical properties, taste and appearance of food.

Physical properties of food

Physical properties describe the **state** of the food material – for instance, whether it is liquid or solid, hot or cold.

Below are useful words for describing the physical properties of food:

liquid, solid, foam, gel	hard, soft	pliable, elastic
hot, cold	runny, viscous	crumbly, brittle
smooth, lumpy	heavy, light	absorbent

The examples in the panel will help you understand what they mean.

Foods have different physical properties

Breakfast cereal with milk
The cereal is solid, hard and crumbly. The milk is liquid, smooth and runny. The cereal will absorb the milk and become pliable. If the milk is from the refrigerator it will be cold.

Spaghetti bolognese
The spaghetti is solid, soft and pliable. The sauce is liquid, lumpy and viscous. The sauce sticks to the spaghetti but is not absorbed. Both are hot.

Steamed sponge pudding with custard
The sponge is a solid, light, soft foam. You can see tiny bubbles in it where gas has been trapped in the mixture as it cooked. The custard is liquid, smooth, viscous and hot. It sticks to the sponge and is absorbed by it.

The appearance of food

Food products are designed to be eaten, but most of us judge food first with our eyes! If it looks good, we are more likely to want to eat it.

You can describe the appearance of food in terms of colour, shape and finish as shown in the pictures.

The taste of food

The taste of food is a complex mixture of the smell, flavour and the 'mouth feel' or texture of the food. The taste buds in the tongue can detect four basic **flavours**:

   sour
   sweet
   salt
   bitter.

You can add words such as 'strong' or 'weak' to describe the strength of the flavour. You can use words such as 'savoury', 'fruity' or 'fishy' to describe a type of flavour.

The 'mouth feel' – the **texture** of the food in the mouth – can be difficult to describe. You can use these words to describe the texture of the food when you first taste it:

   soft, firm, hard
   dry, moist
   crumbly, crunchy, brittle
   thin, creamy, sticky.

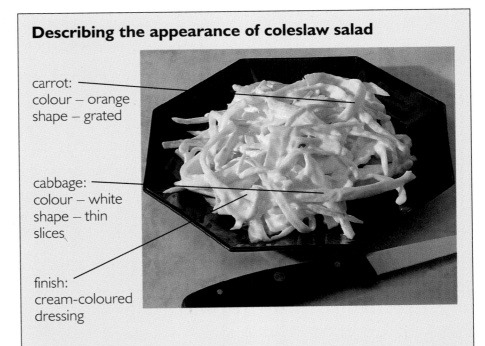

Describing the appearance of coleslaw salad

carrot:
colour – orange
shape – grated

cabbage:
colour – white
shape – thin slices

finish:
cream-coloured dressing

The mix of colours makes the food more attractive.

Describing the appearance of cakes

shape:
round and slightly domed

finish:
white icing with red, yellow and blue decoration

colour:
golden brown

As you chew the food it will change in texture as it mixes with the saliva in your mouth. You may need to use these extra words to describe the texture of the food:

   tough, tender, chewy, rubbery
   gritty, greasy, slimy, gooey.

Resource task

FRT 1

11 DESIGNING AND MAKING WITH FOOD

Food choices

We eat food because we are hungry and because we enjoy eating. Food is so readily available in Britain that we are unlikely to die of starvation. We are more likely to become unhealthy because we eat too much food or too much of a particular type of food.

This section gives information to help you make a healthy choice of food.

Dietary goals

These are what we should aim for in our diet.

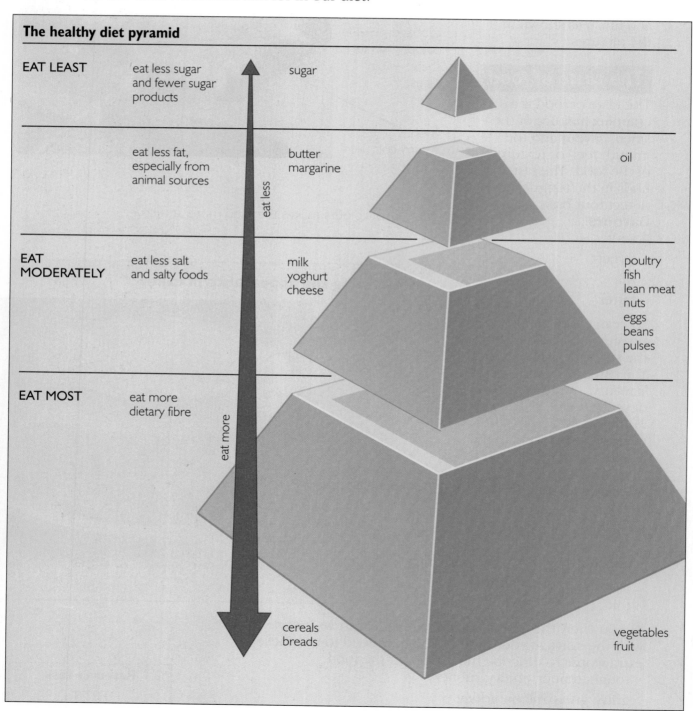

The healthy diet pyramid

EAT LEAST — eat less sugar and fewer sugar products — sugar

eat less fat, especially from animal sources — butter, margarine — oil

EAT MODERATELY — eat less salt and salty foods — milk, yoghurt, cheese — poultry, fish, lean meat, nuts, eggs, beans, pulses

EAT MOST — eat more dietary fibre — cereals, breads — vegetables, fruit

eat less ↑ / eat more ↓

Nutrients in food

Food contains nutrients. These are the chemicals in food which the body uses for growth and repair, energy and warmth and to keep healthy.

By eating a range of foods the body can get all the nutrients it needs. The needs met by the different nutrients are shown in the panel.

Nutrient	Needs met	Foods
Carbohydrates	energy	
Proteins	growth, repair and energy	
Fats	energy	
Vitamins	protection and maintenance of body processes	
Minerals	structure of body and maintenance of body processes	

The body also needs:
- water for all body cells and body processes;
- dietary fibre to help it get rid of waste products.

Resource task
FRT 6

...Food choices

How much do we need?

Nutrients for energy

Carbohydrates, fats and protein provide us with energy. We each have different needs for energy depending on our age and how active we are. So we need different amounts of nutrients to meet our energy requirements.

Our bodies need energy for:
- physical activity like walking, running and jumping;
- bodily processes like the heart beat, breathing and digestion.

The total energy used is called **energy expenditure**. The food we eat should provide enough energy for our expenditure.

> Energy is measured in joules.
> The symbol for joules is J.
> 1 kilojoule (1 kJ) = 1000 joules
> 1 megajoule (1 MJ) = 1000000 joules

Estimated average requirement (EAR) for energy per day		
Age (years)	EAR (MJ/day) ♂	EAR (MJ/day) ♀
7–10	8.24	7.28
11–14	9.27	7.92
15–18	11.51	8.83
19–49	10.60	8.10
50–59	10.60	8.10
60–64	9.93	7.99
65–75	9.71	7.96
75+	8.77	7.61

This person has an energy balance and good health.
energy intake = energy expenditure

This person has an energy imbalance, leading to increase in body fat and poor health.
energy intake > energy expenditure

For an 11–14-year-old, the percentage of EAR supplied by some high-energy foods:

750 kJ — ♂ 8% / ♀ 9.5%
560 kJ — ♂ 6% / ♀ 7%
500 kJ — ♂ 5.5% / ♀ 6.5%
1110 kJ — ♂ 12% / ♀ 14%

Resource task
FRT 7

Other nutrients

The amount of other nutrients needed in the diet also depends on age: growing children usually need more. The figures in the following table are the **reference nutrient intake** (RNI), which is enough for 97 per cent of the population. You can compare a person's nutrient intake with these figures.

Reference nutrient intake (RNI) per day for different age groups								
Age (years)	Protein (g)		Calcium (mg)		Iron (mg)		Vitamin C (mg)	
	♂	♀	♂	♀	♂	♀	♂	♀
7–10	28.3	28.3	550	550	8.7	8.7	30	30
11–14	42.1	42.2	1000	800	11.3	14.8	35	35
15–18	55.2	45.9	1000	800	11.3	14.8	40	40
19–49	55.5	45.0	700	700	8.7	14.8	40	40
50+	55.3	46.5	700	700	8.7	8.7	40	40

Foods high in protein

Foods high in protein and calcium

Foods high in vitamin C

Foods high in iron

Designing food products

Understanding the market

Thinking about what the consumer needs and likes and the properties of the food materials will help you write the specification for the product.

You may need to consider all these points when you write the specification.

▲ *Can you tell who will buy these food products?*

Testing recipes yourself

In developing a food product you might begin by testing an existing recipe. A recipe lists the ingredients used and describes how to combine and cook these to make the food product. Use the ingredients listed and follow the instructions precisely.

Then ask yourself these questions:

The food product itself
- What did I like or dislike about:
  the taste and smell?
  the texture?
  the appearance?

The instructions
- Were they easy to follow?
- Were they always clear?
- Did the ingredients behave in the way described?

The cost
- What does each ingredient cost?
- What is the recipe's total cost?

Your answers to these questions will help you decide whether to use the recipe as it stands in developing a food product or whether it needs modifying.

▲ *Two ways to find out about recipes.* ▼

...Designing food products

Modifying recipes to meet a specification

When you have tested a recipe you may decide to change it. Start by asking yourself:
- Why do I want it to be different?
- What do I want to be different?

This will help you write a specification for the new food product.

Next, think about how you will change the recipe to get the product you want.

Here is an example.

Jamaica Patties

SPECIFICATION
What it has to do:
- taste more spicy
- use different vegetables
- be crunchy in the mouth
What it has to look like:
- golden brown
- a circle folded in half

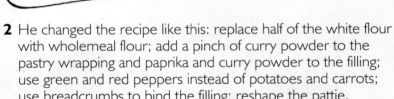

1 Jomo wants to change the recipe for Cornish pasties.

Why? Because his family doesn't like them.

What does he want to be different? The vegetables, the flavour, the colour and the texture.

2 He changed the recipe like this: replace half of the white flour with wholemeal flour; add a pinch of curry powder to the pastry wrapping and paprika and curry powder to the filling; use green and red peppers instead of potatoes and carrots; use breadcrumbs to bind the filling; reshape the pattie.

3 Jomo compared his new product, called Jamaica Patties, with his specification.

4 Then he asked his family what they thought.

You can make changes in four areas

Flavour and texture
You can add ingredients to give a different flavour. Adding dried apricots to a scone recipe makes it more fruity and moist.

You can leave out ingredients to get a different flavour. Leaving out chilli powder in a sauce makes it less spicy.

▲ *Extra ingredients give these biscuits more flavour and different textures.*

The way it is cooked
This will affect the final appearance, flavour, nutritional value and texture.
- Boiling instead of baking prevents browning.
- Grilling instead of frying makes food less greasy.
- Stir-frying instead of boiling keeps vegetables crisp and preserves their flavour and vitamin C.

▲ *The results of changing the method of cooking potatoes.*

Shape and finish
This has a major effect on the final appearance. Sponge cake can be presented as small cakes, as a Swiss roll filled with jam or as a party cake decorated to look like a castle.

▲ *It's the same sponge cake, but what a difference in appearance!*

Nutritional qualities
You can change ingredients to suit particular dietary needs. You might use potatoes and parsnips instead of meat to make a stew suitable for a vegetarian.

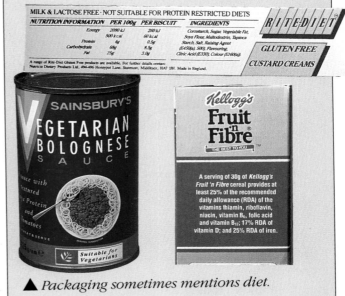

▲ *Packaging sometimes mentions diet.*

You can use **attribute analysis** to help you think up new designs for food products (see page 43).

Resource task
FRT 9

...Designing food products

Choosing recipes to meet a specification

Choosing recipes for wrappings
Many food products enclose or wrap one sort of food in another. The Food Wrappings Chooser Chart shows how food can be wrapped, and the properties and qualities of each type of wrapping. This will help you choose a wrapping recipe to meet your product's specification.

Handling wrappings
The wrapping needs to be strong enough to hold a filling but not so strong that it is difficult to chew. It should not hide the taste of the filling.

You can spoil a wrapping by the way you handle it when you are making it. Follow this advice to get good wrappings.

Stop it sticking
Put some flour on your hands or the rolling-pin, but not too much or you will unbalance the recipe and make the wrapping dry and brittle.

Avoid bursting
Don't put too much filling in the wrapping.

Roll it evenly
Press firmly on the rolling-pin at first and then more gently as you reach the right thickness. Roll it to the shape and size you want.

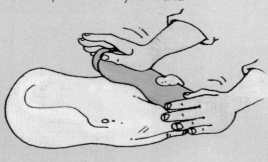

Avoid splitting open
Seal any joints with water or beaten egg.

Keep the structure
Try to get the shape and thickness right first time. If you handle the mixture too much you can destroy the structure.

Flaky and filo pastry lose their flakiness and shortcrust pastry becomes tough.

Resource task

FRT 9

Food Wrappings Chooser Chart

Wrapping	Dip-coat	Wrap-around	Hold	Physical properties	Appearance	Texture
Deep-fried batter	✓	✓		solid and brittle	brown with bubbles on surface; takes shape of food it covers	crisp and crunchy
Shallow-fried batter (pancake)		✓		solid, soft and pliable	cream with brown speckles	soft and chewy
Samosa		✓		solid, brittle when hot, softer when cold	golden brown, some bubbles on surface	crisp when hot
Shortcrust pastry		✓	✓	solid, hard and stiff	golden brown and smooth	crumbly, melts in mouth
Flaky pastry		✓	✓	solid in thin layers and stiff	dark brown, shiny, multi-layered	crisp flakes
Filo pastry		✓	✓	solid in thin layers and brittle	pale brown, darker at edges, multi-layered	crisp flakes

11 DESIGNING AND MAKING WITH FOOD

... Designing food products

Choosing recipes for fillings

Wrappings do not take long to cook, so fillings should be cooked before wrapping or able to be quickly cooked so that they cook inside the wrapping.

Use small pieces of food in the filling to allow the heat to penetrate more easily. Make small pieces by slicing, chopping, mincing or making a purée (pulp).

▲ These can be dipped in coating when raw, then deep fried.

Preparing fillings

Main ingredient	Should I wash it?	Should I cut it up?	Should I pre-cook it?	What other ingredients can I add?
Meat	no	yes	yes	herbs, spices, vegetables, e.g. tomatoes, onions
Fish	yes	yes	yes/no	herbs, spices, vegetables, e.g. mushroom sauce
Cheese	no	yes	no	vegetables, e.g. onions
Fruit	yes	yes	yes/no	sugar, spices
Vegetables	yes	yes/no	yes/no	herbs, spices, sauce

Use egg and milk to hold together or moisten ingredients. Use breadcrumbs to bind together moist or slippery ingredients.

Resource task
FRT 10

Preparing food ingredients

Measuring the quantities needed

If you are preparing a dish like stir-fried vegetables or stew, the exact amount of each ingredient is not important. The ingredients only have to cook, not to rise or set.

For other dishes, you need precise measurements to make the product work. For example, in a sponge mix the proportions of the ingredients affect the rising of the cake. For these, measure the ingredients to be sure they are in the correct proportions. The two main methods of measuring food ingredients are by weight and by volume.

Weighing

You can use kitchen scales for weighing. Most scales measure weight in grams (g) and kilograms (kg). They are accurate to the nearest gram.

If your recipe is in pounds (lbs) and ounces (oz) you should convert them into grams and kilograms. Most recipes use 25 g to 1 oz and 500 g to 1 lb (16 oz).

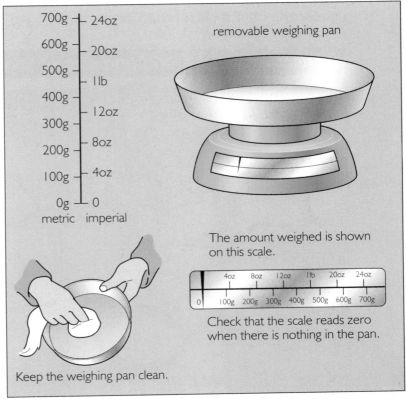

▲ How to use kitchen scales.

Measuring volume

Use a jug to measure large amounts of liquid. Jugs are usually marked both millilitres (ml) and pints. They are only accurate to the nearest 20 ml. For small amounts of liquid or powdery solids you can use measuring spoons. These come in four sizes: 2.5 ml, 5 ml, 10 ml and 15 ml.

Wipe the spoons after use to avoid contaminating one food ingredient with another.

◄ Ways to measure volume.

... Preparing food ingredients

Getting the right shapes and sizes

Once you have decided on the shapes and sizes you need you can choose the right tools for the job.

Controlling the shape and size of food materials

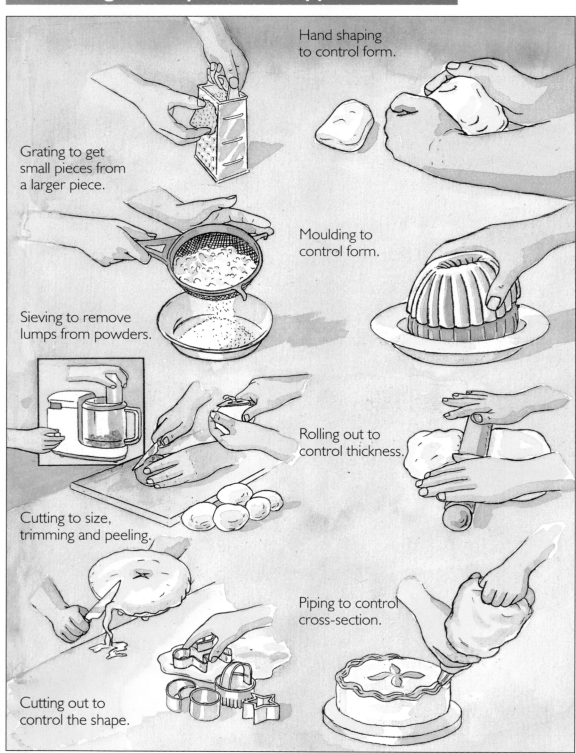

Hand shaping to control form.

Grating to get small pieces from a larger piece.

Moulding to control form.

Sieving to remove lumps from powders.

Rolling out to control thickness.

Cutting to size, trimming and peeling.

Piping to control cross-section.

Cutting out to control the shape.

Combining different food materials

The six main ways of combining food materials are shown in the panel.

Ways to combine food materials

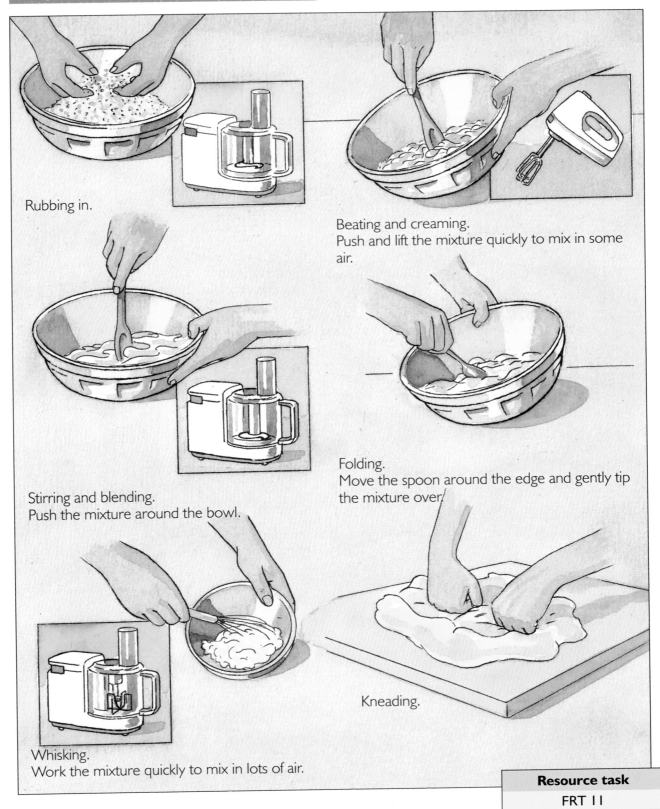

Rubbing in.

Beating and creaming.
Push and lift the mixture quickly to mix in some air.

Stirring and blending.
Push the mixture around the bowl.

Folding.
Move the spoon around the edge and gently tip the mixture over.

Whisking.
Work the mixture quickly to mix in lots of air.

Kneading.

Resource task
FRT 11

Cooking the food

Choosing the equipment and method

Cooking always involves the transfer of heat into food materials. There are many different ways of doing this.

Different methods of cooking

Steaming
The food is in a container with holes in its base over a saucepan of boiling water. The steam surrounds the food and makes it hot. This is a slow method but helps vegetables keep their flavour and vitamin C. It is also used for puddings.

Using a saucepan
The liquid in the saucepan becomes hot through **convection**. The heat is transferred to the surface of the food and is **conducted** slowly to the centre. Use this method for cooking food in small pieces, like vegetables or pasta. It is also used for stews and sauces, when the liquid becomes part of the food.

Using a hot plate, frying-pan or wok
Heat from the hot plate, pan or wok quickly makes the food surface hot. If the food is too thick the outside is burned by the time the inside is cooked. So this method is best for thinly sliced food.

Use a hot plate or pan for food like bacon or sausages or food that forms shallow pools of liquid like eggs or chapatis. Use a wok for stir-frying food divided into small pieces.

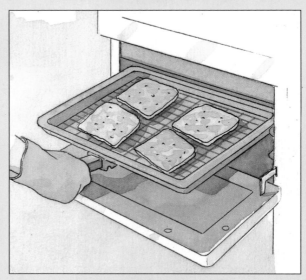

Using the grill or toaster
Radiation makes the food surface hot. This heat is conducted only slowly to the centre of the food. If the food is too thick the outside is burned by the time the inside is cooked. This method is best for food that is in thin slices, such as sausages, bacon or bread.

Using a deep-fat fryer
This works like a saucepan. The food is often held in a wire basket so it can be removed and drained easily. Fat gets much hotter than water, so a crisp coating is formed on the outside of the food and it is browned evenly all over. This method can be used for quick-cooking fish, meat, fruit and vegetables in batter, chips and doughnuts.

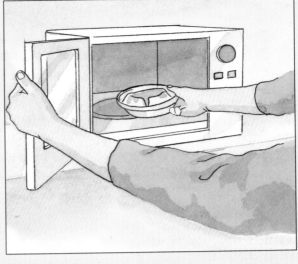

Using a microwave cooker
Microwave radiation makes water molecules in the food vibrate. This fast movement makes the food hot. Food with a high water content heats up quickly.

The microwaves can only penetrate food up to a depth of about 50 mm. This method is particularly useful for defrosting frozen food and heating ready-cooked convenience food.

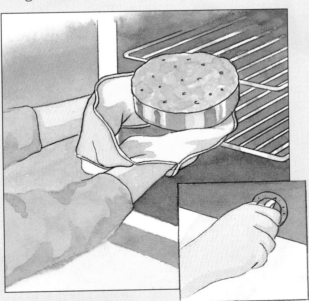

Using an oven
The air inside the oven becomes hot through convection. This heat is transferred to the food's surface. It is conducted slowly to the centre of the food. The oven can be set at a temperature to ensure that the outside of the food does not burn while the inside is cooking.

Use this method for roasting joints of meat and poultry, baking mixtures that have to rise and set, like breads and cakes, and cooking casseroles.

Using a pressure cooker
Steam under pressure can reach temperatures of over 100°C. The super-heated steam surrounds the food and makes it hot. This cooks the food much faster than ordinary steaming.

Resource task
FRT 11

... Cooking the food

What happens to food as it is cooked?

When food is cooked, the flavour, texture and colour change. What changes take place depend on the method of cooking and the ingredients.

Many foods change in several ways when cooked. Cake mixtures rise, set and brown. Biscuits brown but remain soft until they are cool.

Changes that take place when food is cooked

Browning
The surfaces of these foods have become brown and crisp during cooking.

Rising
These foods have risen during cooking.

Setting
These foods have set firm during cooking.

Thickening
This sauce has thickened during cooking.

Is it ready?

It is easy to spoil food by cooking it too much or not enough. Under-cooked food can be dangerous as well as unpleasant because harmful bacteria have not been destroyed in the cooking process. Over-cooked food is often tough or tastes burned.

To tell whether the food you are preparing is properly cooked, ask yourself these questions:

- Is it as brown and crisp as I would like?
- Is it as soft or tender as I would like?
- Is it as set or firm as I would like?
- Is it as thick as I would like?
- Is it cooked through?

Ways of answering these questions are shown in the panel.

Ways to tell if food is cooked

Test for browning by looking, and for crispness by touching gently with a pointed knife.

Test for thickness by seeing how fast it drips off a spoon. A coating sauce should coat the back of a spoon.

Test for tenderness by seeing if a skewer or sharp knife will push in easily.

Gently push a skewer into a fruit cake. If it comes out clean the cake is cooked.

Test for setting by shaking gently. If it is still liquid you will see ripples in the middle.

Use a cooking thermometer to tell if meat is cooked through.

Resource task

FRT 9

Finishing touches

The final appearance of a food product is important. It should be attractive to those who are going to buy or eat it. The two main ways to make food products look better are:
- including decoration as part of the design;
- adding decoration as part of the way it is served.

Here are some examples.

Use a glaze to give the product sheen.

Use coloured ingredients.

Use pastry pieces to decorate pies or flans.

Coat cakes and biscuits with icing.
Shape food so it looks interesting.

Garnish with small vegetables.

Use chocolate pieces, wafers and candied fruit as toppings for ice creams and jellies.

Add a salad.

Sensory evaluation tests

What will other people think about your food product? Use at least ten people for any tests that you carry out for a range of opinions.

Preparing for the test

- Draw up an answer sheet for the tasting panel.
- Give each taster a glass of water to sip between samples to take away the taste.
- Use the same type of plate or container for each sample.
- Label the samples of food so that the taster does not learn anything from the label. Symbols are best, such as ▲, ■, ●, or random numbers such as 369 or 472. Do not use 1, 2, 3 or A, B, C, as they may give tasters the impression that number 1 or letter A is best.
- Hold the tasting session somewhere not affected by cooking smells.
- Tell the tasters not to talk to one another. It is their individual opinions that you want.
- Give each person the same size sample to taste – enough for about two bites or sips.

The three types of test you can use are:
Ranking tests
Preference tests
Difference tests

Ranking tests

This type of test will help evaluate the strength of a particular quality of the food. It is good for decisions on flavour, colour and texture. Here is an example.

Which dried fruit gives the most moist biscuit?

Make three batches of the biscuits, each with the same amount of a different dried fruit. Set up your tasting panel like this:

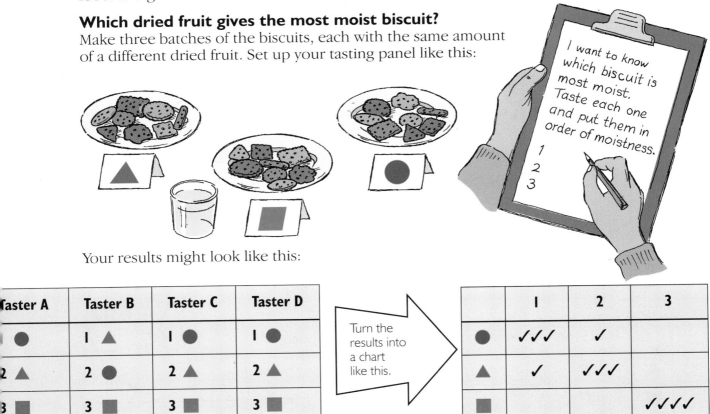

Your results might look like this:

Taster A	Taster B	Taster C	Taster D
1 ●	1 ▲	1 ●	1 ●
2 ▲	2 ●	2 ▲	2 ▲
3 ■	3 ■	3 ■	3 ■

Turn the results into a chart like this.

	1	2	3
●	✓✓✓	✓	
▲	✓	✓✓	
■			✓✓✓✓

The ticks tell you the **rank order**. From this small sample, biscuit ● is thought to be the most moist.

Note: the ranking test does not tell you whether the panel preferred one. For that you need a preference test.

Resource task

FRT 3

...Sensory evaluation tests

Preference test

This type of test is used to find out how much a person likes or dislikes a food.

Use a five-point scale of descriptive words or faces (for young children), like the one here, to help people describe how much they like a product.

1 Like very much   **2** Like moderately   **3** Neither like nor dislike   **4** Dislike moderately   **5** Dislike very much

How much do you like the cakes?
Set up a tasting panel like this:

Preference test
1. Taste a sample
2. Circle the number which best describes how much you like or dislike the food
3. Take a drink of water to clean your palate
4. Taste the next sample. Circle the number which best describes how much you like or dislike the food. Do not compare the samples
Repeat this until you have tested all the food samples

	😊	🙂	😐	🙁	☹
▲	1	2	3	4	5
■	1	2	3	4	5
●	1	2	3	4	5

With ten tasters your results might look like this:

▲	5	4	5	4	4	2	4	3	5	3
■	3	3	5	4	2	1	3	4	2	5
●	1	1	1	3	2	3	1	2	3	2

	Total score	**Average score**	**Conclusion**
▲	39	39/10 = 3.9	disliked moderately
■	32	32/10 = 3.2	neither liked nor disliked
●	19	19/10 = 1.9	liked moderately

Turn the results into a chart like this.

You can draw a conclusion about how much the tasting panel liked each sample as shown in the last column of the table. This type of test is called a **hedonic ranking** test.

Resource task
FRT 4

Difference test

This test is useful for finding out whether people can tell the difference between slightly different food products. Here is an example.

Can people tell the difference between a fruit dessert made with sugar and one made with artificial sweetener?

Make up the two recipes and prepare three samples of food labelled with symbols:

- sample ▲ made with sugar;
- sample ● made with artificial sweetener;
- sample ■ made with artificial sweetener.

Samples ● and ■ are identical.

Give each taster an answer sheet like the one shown.

You can then count up the number who could tell the difference.

It is probable that the correct answer chosen by chance is 33.3 per cent or one third. If more than one third of your tasters choose sample ▲ as being different, you need to make more changes. If fewer than one third choose sample ▲ as being different, then your recipe is acceptable.

This is often called a **triangle test**. Can you explain why?

Being a food product tester

Sometimes you will have to test a food product on your own. You might not be sure that the product is good enough yet to set up a tasting panel. Or, there may not be enough time for a tasting panel. Use your own judgement. Here's how to do it.

Look at it carefully.   Sniff it carefully.   Taste a tiny piece.

What can I do to improve the product?

Chew a small piece.   Keep chewing.

Resource task
FRT 5

11 DESIGNING AND MAKING WITH FOOD

Prolonging shelf-life

▲ Rotten food usually looks bad but eggs that cause food poisoning look OK.

Keeping food fresh

You can keep food in good condition by following these rules:

- check the 'use by' or 'best before' dates on the label;
- follow any storage instructions given;
- keep food covered or wrapped to stop it drying out;
- keep foods cool. This slows down the **enzyme** action in raw food that causes it to spoil. It will also slow down the growth of micro-organisms that make food go bad.

Shelf-life is the length of time that a food, if stored correctly, will remain safe to eat and keep its physical properties, appearance and taste. If it is stored for longer than its shelf-life or incorrectly it will spoil.

What makes food go bad?

Micro-organisms are yeasts, moulds and bacteria that grow in foods and make it go bad. Food that has been attacked by micro-organisms often smells 'off' or looks 'mouldy'.

The yeast in fruit-flavoured yoghurt can make the yoghurt **ferment**, producing a gas which pushes the lid up on the yoghurt pot. The yoghurt has an 'off' flavour.

Some bacteria make food look 'slimy'.

Bacteria such as salmonella are particularly dangerous because the food does not look bad. The taste and appearance are not affected but the bacteria cause food poisoning.

Laboratory tests find the cause of food poisoning outbreaks. Salmonella are destroyed by high temperatures during cooking.

Resource task
FRT 8

Preserving food

People have always tried to preserve food for the winter months when supplies are scarce. Our full supermarket shelves shows how successful we are today in preserving foods.

Food is preserved by killing or preventing the growth of the micro-organisms that spoil it. Micro-organisms need:

- moisture;
- neutral surroundings, neither **acid** nor **alkaline**;
- temperatures of between 5°C and 63°C.

By controlling conditions so that micro-organisms cannot live, you can preserve food. The preservation may alter physical properties, appearance and taste.

Canning and bottling
The high temperatures used in processing food destroy micro-organisms but change its physical properties, appearance and taste.

Drying
This destroys micro-organisms by removing moisture from food but changes its physical properties, appearance and taste.

Irradiation
This destroys the micro-organisms but has little effect on the food's physical properties, appearance and taste.

Pickling
This makes the surroundings of the food **acid**, and changes its physical properties, appearance and taste.

Freezing
This does not destroy micro-organisms but stops them reproducing. It has little effect on the food's physical properties, appearance and taste.

Careful practice and food safety

You must work safely and **hygienically** when you design and make with food. Here are some important rules to follow in preparing food:

Personal readiness
- Before you start wash your hands in hot, soapy water in a hand-basin and dry with a hand-towel or hot-air drier;
- Cover cuts with a clear waterproof dressing;
- Wear clean protective clothing;
- Tie back or cover hair;
- Remove jewellery.

The materials you intend to use
- Store food in pest-proof containers;
- Handle as little as possible;
- Keep raw food and cooked foods separate;
- Make sure that frozen meat is thawed completely before cooking;
- Prevent bacteria multiplying by cooling and storing food below 5°C or above 63°C if it is being kept hot;
- Do not lick fingers or put tasting spoons into food.

The place where you work
- Keep your work area tidy;
- Clean equipment, work surfaces, sinks and cookers with hot, soapy water;
- Clear away waste food by wrapping and putting into bins;
- Empty bins regularly;
- Keep floors clean.

The equipment you intend to use
- Take extra care with sharp knives and graters;
- Make sure that pans of hot liquid do not get knocked over;
- Turn ovens and heating rings off after use;
- Use blenders and food mixers very carefully.

Glossary

This list explains how some words are used in design and technology. Some words may have different meanings in other subjects. A **bold** word in an explanation means that the word has its own entry in this glossary.

ABS (acrylonitrile butadiene styrene) A stiff, strong, tough plastic useful for making frameworks and mechanical parts.

abrasive A material which smooths and removes marks from wood, plastics and metals by scraping and grinding.

abutment The end supports of a bridge which are embedded in the ground and support the bridge by resisting forces exerted on them by the bridge.

acid Any substance with a pH less than 7 and turns litmus red.

adhesives Substances used to stick materials together.

aesthetics The area of design concerned with making products look attractive.

alkali Any substance with a pH greater than 7 and turns litmus blue.

amplifier The name used for any **component** that turns a small **input** into a large **output** e.g. a **transistor** which turns a small input current into a larger output current.

analogue input Signals received by a computer **interface** box from electronic **sensors** that sense constantly changing values such as temperature, light or sound levels. These signals have to be converted into **digital** form in the interface box before they can be processed by the computer.

anneal To heat and then cool metal so that it is easier to shape.

annotate Add brief notes to your design sketches to make things clearer or to give more detail.

anthropometric data Information about people's shapes and sizes.

assembly The way the parts of a product are fitted together.

attribute analysis A way of describing a product so that you can develop new designs for that product.

axle The **shaft** on which wheels are carried. The wheels are either fixed so that they turn with the axle, or able to spin freely on the axle.

base One of the legs on a **transistor** through which a small **input** current flows.

bearings Material between a rotating **shaft** and its support which reduces friction e.g. ball bearings and roller bearings.

bell crank A right-angled **lever** used to change the direction of **linear movement**.

belt See **pulley**.

bevel gears Gears with sloping sides that are used to transmit motion between **shafts** that are at an angle to each other.

bimetallic strip A strip made of two metals that curls when it is heated.

biodegradable Able to be broken down by the action of micro-organisms.

brainstorming A way for a group of people to think of lots of ideas quickly.

CAD (computer assisted design) Using a computer to design a product or part of a product by drawing it on-screen.

CAM (computer aided manufacture) Using a computer to control the machine tools that make a design.

cam a non-circular wheel that rotates and moves a **follower**. It can convert **rotary movement** into **reciprocating** or **oscillating movement**.

cantilever A beam that is supported at one end only.

carbon dioxide A colourless, odourless gas present in the atmosphere. It is formed during respiration and when organic materials are burnt or decompose.

carcass A box construction which is made from sides joined together.

casing A channel made in fabric for tape, cord or elastic.

cast A product is cast by pouring a liquid into a **mould**, leaving it to set solid and then removing the mould.

casting The name given to a product that has been **cast**.

changeover switch A switch that will turn off one switch at the same time as turning on another.

circuit An arrangement of **components** that provides a continuous pathway through which an electric current can flow.

cleat A mechanism which uses **cams** to hold ropes tightly on sailing boats.

closed-loop system A **system** that responds to changes in its **environment** by using **feedback** from within the system.

closed brief See **design brief**.

closed question A question to which the answer is a fact rather than a matter of opinion.

cm (centimetre) A unit for measuring length. 100 cm = 1 metre (m).

collector One of the legs on a **transistor** through which a large **output** current flows.

components The name given to the parts that make up a product.

compression When a structure is being squeezed and it pushes back in the opposite direction it is in compression.

conduction The way heat travels through a solid by particles vibrating and passing this vibration on to those particles next to them.

conductor A conductor is a material that allows something to pass through it. An **electrical conductor** allows electricity to pass through it. A **thermal conductor** allows heat to pass through it.

consistency When you are describing food this means the thickness of a mixture.

convection The way that heat travels in liquids and gases by currents in which warm material moves upwards and cold material moves downwards.

convenience foods Food products that are easy to prepare for eating.

conventions Ways of showing information on **working drawings** that are recognized and understood all over the world.

corroded Materials, usually metal or stone, that have been eaten away by the action of their **environment**.

couplings Connectors for **shafts** which transmit movement from one shaft to another.

cramp An adjustable tool for holding materials steady while you work on them or while **adhesives** set.

GLOSSARY

crank An arm fixed to a **shaft** for transmitting motion from that shaft. When the shaft rotates the arm rotates too.

cut-away view A drawing which shows the construction and internal detail of an object by showing the outer parts 'cut away'.

Darlington pair A special arrangement of two **transistors** in an electronic **circuit**, which gives a more sensitive and faster response to a **sensor** than a single transistor.

demonstration model A model of a product or **system** which is used to show how the product or **system** works.

dense Closely packed. A dense material is heavy for its size, e.g. lead.

design brief A summary of the aims of your design and the kind of product that is needed. A **closed brief** says what the product will be. An **open brief** leaves it to you to decide.

development (net) A flat sheet of material (e.g. paper or card) shaped and cut so that it can be folded to make a 3D form.

digital input Signals received by a computer **interface** box from switches that are either on or off.

diode An electronic **circuit component** which can be used with a **relay** to protect the circuit from damage.

drafting patterns Preparing an accurate paper or card pattern for a textile product, including all the instructions for making.

DTP (Desk-top publishing) Using a computer to design the layout of pages for a broadsheet, newspaper, magazine or book.

EAR (estimated average requirement) The amount of energy an average person needs each day.

eccentric A circular wheel with an off-centre **axle**.

electronic control system The electronic **circuits** used to control a device, e.g. the circuits controlling the opening and closing of automatic doors.

elevation Vertical square-on views of an object or building, to show detail for making, usually including front elevation and end elevation.

emitter One of the legs of a **transistor** through which a large **output** current flows.

emotion Feelings (not thoughts and ideas).

energy expenditure The total amount of energy that a person uses during a day. It will depend on what they do and for how long.

environment The surroundings, e.g. a room, a town, a park, a forest, etc.

enzyme A biological catalyst. (A catalyst is a substance that increases the rate of a chemical reaction.)

ergonome A scale model of a person to help in working out the **ergonomics** of your design.

ergonomics The study of how easy it is for people to use their working **environment**.

exploded view A drawing which shows how all the parts of an object fit together by showing the object pulled apart.

eye level A line drawn to represent the horizon in a **perspective drawing**.

feedback Information which tells a **closed-loop system** when something in its **environment** has changed.

ferment A food material ferments when micro-organisms break down the material producing **carbon dioxide** and sometimes alcohol.

fittings Things used to fix materials together, e.g. nails, screws, nuts and bolts.

flavour The sensations detected by the tongue which, with smell and **texture**, give food its taste.

flow chart A way of planning how to carry out a task by drawing a sequence of boxes joined by arrows. Each box contains a short statement about one stage.

flux A substance used to keep metals clean during soldering.

follower Usually a **slider** or **lever** that is moved by a **cam** or **eccentric**.

font A particular style or design of lettering.

four-bar linkage A mechanism used to turn **rotary movement** into **oscillating movement**.

fulcrum (pivot) The point of support of a **lever**, around which it moves.

Gantt chart A chart for planning complex operations. It shows each stage, the time it will take and where it fits in the overall sequence.

gear A toothed wheel, usually fixed to a **shaft** so that it rotates at the same speed as the shaft.

gearing down Using a **gear train** to increase **output** force and reduce speed.

gearing up Using a **gear train** to decrease **output** force and increase speed.

gear train Gear wheels with teeth meshed together so that one drives the other.

graphics Writing, drawing, printing, decorating, etc. on a flat surface.

grid A sheet marked out with columns and lines to help designers keep the same basic look for each page of a publication. Or squared paper useful for drawing maps, plans and 3D views, enlarging and reducing pictures, developing repeat patterns, **drafting patterns** and planning **circuits** and room layouts.

hedonic ranking Arranging something in order from 'like most' to 'like least' (hedonic is a Greek word meaning pleasure).

hydraulic Operated by liquid under pressure.

hygienic Working in a clean and therefore healthy way.

idler gear A **gear** wheel placed between two others in a **gear train** so that the outer wheels both turn in the same direction.

indicator A light that comes on to tell you that something is happening. Often these are **LEDs**.

input All the things that go into a **system** are called inputs.

insulation A material that is used to prevent heat transfer. It will keep cold things cold and hot things hot.

intellectual Thoughts and ideas (not **emotion**).

interface The place of contact between objects, people and machines, e.g. a computer interface.

interviewee A person being interviewed.

isometric view A way of showing three dimensions on a drawing. Special isometric **grid** paper can be used.

J (joule) A unit for measuring energy.

jig A mechanism for holding the object you are working and guiding a tool.

kJ (kilojoule) A unit for measuring energy. 1 kJ = 1000 joules (J).

latch A way of keeping the **output** on when the **input** has stopped, e.g. the wiring of a **relay** to bypass a **transistor** in a sensing **circuit**.

LDR (light-dependent resistor) A **sensor** that can be used to detect changes in light level. It has a high **resistance** in the dark and a low resistance in the light.

leaf springs Flat springs which try to straighten when they are bent.

LED (light-emitting diode) An electric **circuit component** that lights up when a current passes through it.

GLOSSARY

levers Bars or rods that move about a **pivot** or **fulcrum**.

linear movement Movement in a straight line.

linkages Parts used to connect the **components** of a mechanism so that they move in a particular way, e.g. **crank**, link and **slider**.

m (metre) A unit for measuring length. 1m = 100 centimetres (cm).

mdf (medium density fibreboard) A board made from compacted wood fibres.

microprocessor A large integrated circuit that is programmed to control electrical machines.

microswitch A small switch that can be operated by weak forces.

microwaves Radiation that can be used to defrost and cook food.

MJ (megajoule) A unit for measuring energy. 1 MJ = 1 000 000 joules (J).

ml (millilitre) A unit for measuring volume of liquids. 1000 ml = 1 litre (l).

mm (millimetre) A unit for measuring length. 10 mm = 1 centimetre (cm).

mould A hollow shape into which a liquid is poured and left to set solid to produce a product in the shape of the mould.

N (newton) A unit for measuring force.

net (development) A flat sheet of material such as paper or card, shaped and cut so that it can be folded to make a 3D form.

nutrition To do with eating and what we need to eat to keep fit and healthy.

oblique view A simple way of drawing objects to look 3D based on a flat front or side view.

ohm (Ω) A unit for measuring **resistance** in electric circuits.

one-point perspective drawing A way of making a perspective drawing by using the illusion that things converge in the distance to a single **vanishing point**.

on/off switch A switch that stays in the position set, either on or off.

opaque An opaque material is one that you cannot see though.

open brief See **design brief**.

open-ended question A question to which there is more than one right answer. The answer is an opinion.

open-loop system A **system** without **feedback** which cannot respond to changes in its **environment**.

operator interface The parts of a **system** that an operator looks at, touches, talks to or handles.

orthographic projection A way of drawing the detail of a 3D object or building in 2D by showing square-on views of the different sides.

oscillating movement Swinging to and fro, like a pendulum.

output All the things that come out of a system are called outputs.

overlocking machine stitching used to neaten the edges of textile items and prevent fraying.

parallel linkage A mechanism used to keep moving parts parallel.

PCB (printed circuit board) A board with copper tracks that connect electrical and electronic **components** as shown on a **circuit** diagram.

peg and slot mechanism A mechanism in which a peg on a rotating wheel causes a slotted **follower** to move backwards and forwards.

performance specification A description of what the product you design will have to do and possibly how it should look, how it should work and any other requirements it must meet.

perspective drawing A way of drawing that shows depth as well as height and length.

photosynthesis The process by which living green plants use energy from sunlight to produce food (glucose) from **carbon dioxide** and water.

Pie chart A circular chart showing data in the form of a circle divided up into portions, like the slices in a pie.

pin joints A method of joining **linkages**, which allows the free movement of the linkages about the joint.

pivot (fulcrum) The point of support of a **lever,** around which it moves.

pneumatic Driven by compressed air.

positive drive Force being transmitted without slipping by a chain and **sprocket** or toothed **belt** and **pulley**.

potential divider An arrangement of two **resistors** used to divide the voltage in a **circuit**.

presentation drawing/model A well-finished drawing/model of your design idea to show the client how it will look or work.

procedure The steps for performing a task.

program The set of instructions used by a computer to carry out a task.

properties Special features of a material that make it useful for a particular job.

prototype An accurate detailed model showing what a design will look like and, sometimes, how it will work.

pulley A wheel used with a **belt** to change **rotary movement** to **linear movement**.

push-to-break switch A switch that will break a connection when pushed but returns to the on position when released.

push-to-make switch A switch that will make a connection when pushed but returns to the off position when released.

PVC (polyvinyl chloride) A plastic material which comes in two forms. The rigid form is useful for containers. The pliable form is used for sheeting and as a fabric.

quenching To cool suddenly by plunging into water or oil.

rack and pinion Gear system for changing rotary movement to **linear movement**.

radiation The way heat is transmitted across empty space.

rank order A sequence, the order of which is worked out according to the strength or size of a particular feature.

recipe A list of all the ingredients needed for a food product and step-by-step instructions for making it.

reciprocating movement Movement backwards and forwards in a straight line.

relay An electrical **component** that uses an electromagnet powered by a small current from one circuit to operate switches in a separate **circuit** usually connected to a different power supply.

resistance In a structure this is the force provided by the structure in response to the load on the structure. In an electric **circuit** this is a measure of the difficulty the electric current has in flowing through the circuit.

resistor A resistor is a **component** in an electric **circuit** that restricts the flow of current by providing a **resistance** to it.

rocking arm A **lever** that is made to rock to and fro, or oscillate, by a rotating **crank** and connecting **link**.

RNI (reference nutrient intake) The amount of nutrients needed by 97 per cent of the people within a particular age group.

rotary movement Movement in a complete circle.

r.p.m. Revolutions per minute.

section drawing A way of showing detail hidden inside an object by drawing what you would see if you cut through it.

see-through drawing A way of showing detail hidden inside an object by drawing it as if you could see through its 'skin'.

Glossary

selvedge The non-fraying edge of woven fabric, parallel to the **warp**.

sensor An electrical circuit **component** that can detect changes in its surroundings.

shaft A rod which transmits **rotary movement** along its length.

shear Cutting thin sheet material with tools that work like scissors, e.g. tin snips.

single-point perspective view A way of drawing scenes in perspective based on the idea that everything disappears at a single point on the horizon. Special **grid** paper can be used.

sketch model 3D model made using quick techniques and easy-to-work materials so that you can see whether an idea will work and how it will look.

slider A **lever** that is made to slide backwards and forwards in a straight line by a rotating **crank** and connecting link.

social Relationships with other people.

sprocket A toothed wheel used with a **chain** to change **rotary movement** to **linear movement**.

stable A stable structure is not easily overbalanced.

state The form which a material can take – solid, liquid or gas.

storyboard A way of planning how to carry out a task by drawing a sequence of pictures.

straight grain See **warp**.

strategies The different ways you can use to tackle a task.

subsystems Smaller units into which **systems** can be broken down so that they are easier to understand.

surface mount Displaying 2D work by fixing it onto sheets of mounting paper or card.

swatch A small sample or collection of samples of fabric.

system A collection of objects and/or people that work together to do a job. Thinking about problems in terms of systems helps with the design of complicated products.

system boundary This contains all the objects and/or people that you decide to include in your **system**.

system diagram A diagram showing how **subsystems** connect together to form complete **systems**.

synthetic A material that has been manufactured.

target chart A way of evaluating a product by listing winners and losers on a circular chart.

tempering Reheating tool steel very gently to particular temperatures to obtain the degree of toughness and hardness needed for different types of tools. The colour of the steel shows the temperature reached.

tension When a structure, a rope or a spring is being stretched and pulls back in the other direction, it is in tension.

texture The feel of food as it is chewed in the mouth.

thermal To do with heat.

thermistor A **sensor** that can be used to detect changes in temperature as part of an electronic **circuit**.

thermoplastic A rigid material that becomes soft and pliable on heating and rigid again on cooling.

thermosetting A property of some liquids that set solid and then cannot be made to melt or soften by heating.

thumbnails Quick sketches, without detail, for developing your design ideas on paper.

tolerance Allowed variation in size, weight, etc. of the parts for a product.

tone Areas of light and shade added to an object to make it look solid.

torsion spring A spring which tries to close when opened out.

transistor An electronic **component** with three legs. It is turned on by a small current entering the **base** leg which then allows a larger current to flow through the other two legs – the **collector** and **emitter**.

transmission mechanisms Mechanisms that transmit force or movement from one place to another.

transparent A transparent material is one that you can see through. A translucent material will let light through but it cannot be seen through clearly.

triangle test A way of finding out whether people can tell the difference between slightly different things by giving them three samples to test.

trim Tidying up material after cutting it roughly to shape to get exactly the shape you want.

typeface See **font**.

user guide A set of instructions supplied with a product to ensure that it is used effectively and safely.

user interface The parts of a **system** that a user sees, touches, talks to or handles.

user trip Evaluating a product by trying it out.

UV Ultraviolet light.

V (volt) A unit for measuring the force pushing an electric current around a **circuit**.

vanishing point The point at which lines converge in **perspective drawing**.

warp (straight grain) The lengthwise threads in woven fabric, parallel to the **selvedge**.

weft The threads woven across the width of the fabric through the lengthwise **warp** threads.

window mount Displaying 2D work by fixing it behind a mounting sheet of card or paper with a hole or 'window' cut in it.

working drawing A detailed drawing containing all the information needed to make your product.

yarn Thread produced by spinning – twisting – natural or **synthetic** fibres. Yarn is combined by weaving or knitting to form fabrics.

Index

A

3D form 47, 50–1, 54, 63, 79
abrasives 210
adhesives 204
advertising 2, 84
alarms 78, 155, 157, 170–1, 176
  intruder 24, 39
appearance 8–11, 62–3, 124, 184, 209–10, 219
assembly details 95, 97
attribute analysis 43
axis of rotation 100, 105, 107
axles 61, 100, 102

B

bags 20–1, 38, 142–3
batteries 152, 154, 160, 173
bearings 102, 103
bicycles 22, 78, 101, 103, 110
bimetallic strip 163, 172
briefs 36–7
buildings 5, 24, 28, 46, 60, 63

C

cams 114–16
casing 134
chain drive 107–9
charts
  bar 90
  bubble 41
  flow 70, 175
  Gantt 71
  mechanism movement 118–19
  pie 91
  target 74
chooser charts
  adhesives 204
  electric components 160–1
  electronics 172–3
  fabric decoration 137
  fabrics 124
  fastenings 135
  finishes 213
  fittings 205
  food wrappings 229
  metals 188
  plastics 189
  resistant materials 186–7
  tools 214–15
circuit diagrams 156–9, 160–1, 163–9, 173
clothing
  design 5, 9–10
  fabrics 125–8
  from rectangles 144–5
  modelling for fit 59
  paper patterns 59, 92
  performance testing 73
  textile designs 82–3
  warmth 12–13
components 95, 154, 158, 160, 169
compression 14, 120, 178, 180
computers 62–5, 88–9, 91, 174–7, 203
connections 42, 153
construction 54–5, 129–35, 183
control systems 162, 165, 174–7
conventions 94
cooking 216, 218, 227
  finishing 238
  methods 234–5
  processes 236–7
  tools 232, 244
cost 43, 75, 225
cranks 103, 110–13
crating 48, 79
cultures 2, 10, 33

D

data, recording 31, 41
decoration 62, 80–1, 136–7
design
  briefs 36–7
  food products 224–31
  ideas 40–4, 76–98
  models 44–65
  transfer 192–3
  working methods 30
desk-top publishing 63, 89
developments, modelling 59
diagrams 52–5, 70, 77, 97
direction 99, 104, 107, 111, 119, 121–2
displays 87
document presentation 63, 88–91
drawings 28, 42, 47, 54, 79, 93
  see also sketches
  background 85
  computer models 62
  exploded view 55, 95, 97, 171
  perspective 86
  presentation 77–8
  product modelling 52–5
  surface texture 62, 80–1
  techniques 48–9
  working 77, 94
drilling 197–8
drink 2, 13, 18–19, 34

E

eccentrics 114–16
edges 128, 132–3, 192
electric circuits 152–61
electric motors 154, 157, 159–60
electrical conductivity 184
electricity 6, 23
electronic circuits 46, 61, 64, 162–73
elevations 93
energy 6–7, 64, 120, 221, 222
environment 5–7, 28–9, 60, 63, 75, 186
ergonomics 4, 56, 65
evaluating 72–5, 98, 239–41
exhibitions 87

F

fabrics 123–51
fashion 8, 33, 82–3
fastenings 135, 143
feedback 69
fibres 123, 147, 149, 150
finish 210–13, 227, 238
fit 59, 142, 147, 149, 150
fittings 205, 208
flavour 219, 225, 227, 241
flux 153, 207
food 2, 16–17, 216–44
  choice 220
  colour 34
  combining 233
  cooking 227, 234–7
  design 39–41, 224–31
  evaluating 239–41
  fillings 230
  fresh 216, 242
  ingredients 231–3
  pre-cooked 217
  presenting ideas 84–5
  preserving 242–3
  properties 218–19
  quantities 231
  recipes 92, 225–33
  taste 219
  testing 239–41
  wrappings 228–9
force 178–83
  applied/maintained 101, 115, 117, 120, 122
  decrease 100, 106, 109, 113, 119
  increase 100, 102, 105, 108, 112, 119
  transmission 101, 103, 109, 113, 122
forming 200–1
frameworks 208
fraying 130, 132–3
friction 102
fulcrum 110, 111, 113

G

games 23, 25–7
gears 61, 100, 103, 104–6, 109
graphics 90, 97
graphs 90
grids 46, 79, 89
guidelines 45, 48

H

handles 4, 56, 143
hats 73–5, 150–1
heating 13, 66, 69
highlights 80–1
hollowing 200–1
hydraulic systems 61
hygiene 244

INDEX

I

ideas
  brainstorming 40–1
  communicating 76–98
  in context 85–6
  developing 76
  generating 40–3
  modelling 44, 65
  presenting 80–91
  selling 77
  sketches 42
  visualizing 76
image boards 35, 173
information 32, 92–5
inputs 67, 104, 162, 170, 172, 175, 177
instructions 97
insulation 13, 64, 125
interfaces 68, 171, 174
interior design 5, 28–9, 53
  modelling 46, 60, 63

J

jewellery 9–10, 38, 42, 50
joining 103, 110, 129–35, 204–7

L

LDR see light-dependent resistors
LED see light-emitting diodes
levers 100, 101, 110–13, 114
light bulbs 154, 158, 160
light and shade 49, 81
light-dependent resistor (LDR) 163, 165–6, 172
light-emitting diodes (LEDs) 153, 154
lighting control system 165–6
likes see preferences
linkages 110–13
loads 178–83
losers 74

M

maps 46, 53
materials
  attributes 43
  biodegradable 184
  modelling 50–1, 57, 58
  resistant 184–215
measuring 139, 231
mechanisms 61, 64, 99–122
metal 188, 196, 199, 201
  bending/folding 199

brazing 207
casting 202
cleaning 210
corrosion 184
dip-coating 212, 213
fittings 205
hardening/tempering 209
microprocessor 174
modelling 44–65
  3D 47, 50–1, 54, 58–9, 63
  computers 62–5
  developments/nets 59
  grids 46
  kits 61
  materials 50–1, 57, 58
  movement 57
  product performance 56–61
  sketching/diagrams 52–3, 56
  structures 58
models 77
  block 60, 87
  demonstration 96
  full-scale 60
  presentation 87
  prototype 87
  sketch 50, 56
  toys 27
moisture sensor 163, 171, 172
moulding 201
mounts 87
movement
  changing type 102, 104, 107, 117, 118
  direction 99, 104, 107, 111, 119, 121–2
  linear 99, 102, 104, 107, 117
  modelling 57
  oscillating 99, 110, 114
  reciprocating 99, 111, 114, 115
  rotary 99–117
  sensor 163, 172
  transmission 101, 103, 109, 113, 122

N

natural resources 6–7, 75, 186
needs 2–3, 15, 31–5, 75
nets, modelling 59
nutrition 65, 84, 220–3, 227

O

outcomes, evaluating 72–5
outputs 67, 104, 162, 170, 172

P

packaging 7, 35, 51, 59, 84
painting 136, 137, 211, 213
parts 95, 208
patterns 10
  altering 151
  drafting 138–9
  grids 46
  marking out 192–3
  paper 59, 92, 128, 131
pegs and slots 114–16
performance
  business 65
  modelling 52–61, 64–5
  specification 38–9
  testing 73
perspective 86, 93
PIES checklist 33
plans 28, 46, 53, 70–1, 93
plastics 43, 184, 196, 199
  casting 202
  cleaning 210
  fittings 205
  models 51
  types 189
playgrounds 27
power supply 152, 154, 160, 172–3
preferences 2–3, 31–5, 240
presentation drawings 77, 82–5
printed circuit board (PCB) 165, 168–9, 173
printing 88, 136, 137
procedures 176–7
processors 163–6, 170, 172, 175
products
  appearance 62–3
  assembly 95, 97
  casing/style 171, 173
  evaluation 72–5, 98
  performance 38–9, 56–61, 64–5
  safety 23
programs 62–3, 175–6
properties 43, 125, 184–5, 218–19
protein 221–3
pulleys 61, 99, 107–9
puppets 26, 140

R

rack and pinion 104
recreation 25–7
recycling 7, 186
relays 167
renewable energy 7, 75, 96
resistance 155, 164, 178, 181–2
resistant materials 184–215
resistors 154–5, 164, 166
Resource Tasks 1
reversing switch 157, 159, 161

S

safety 22–4, 152–3, 169, 190–1, 195, 244
sanding 197, 210
sawing 194, 198
seams 128, 130–1
sensors 163–6, 172, 176
sewing 103, 111, 114, 129–35
shafts 103
shape 4, 43, 45
  3D form 47, 50–1, 54, 63
  computer modelling 62
  design transfer 192–3
  food products 227
  isometric views 79
shearing 195
sheet materials 200–1
shelter 2, 14–15
shoes 5, 8
shorts 146–7
sketches
  see also drawings
  annotated 45
  cut-away views 55
  ideas 42
  magnified details 54
  modelling 52–3, 56
  rough 76
  sketch model 50, 56
  thumbnail 52, 76
slider 111, 114
slots 114–16
software, modelling 62–3
solder 153, 160, 207
sound, electrical 155, 160
specifications 38–9, 158, 224, 226–7
speed
  decrease 100, 102, 105, 108, 112, 119
  increase 100, 106, 109, 113, 119, 122

springs 101, 120
sprockets 107–9
squeezing 14, 178, 180
stability 58, 181
stiffness 58, 143, 184
stitching 131–2
storyboards 70
strategies 30–75
strength 58, 125, 182–3, 184
structures 27, 58, 61, 178–83
style 8–11, 33
subsystems 67
switches 156–7, 161, 172, 176–7
syringes 121–2
systems 66–9

T
T-shirts 148–9
taste 219, 225
tension 14, 120, 179–80, 183
testing 44, 73, 225, 239–41
textiles 82–3, 123–51
texture 62, 80–1, 219, 225, 227
thermal conductivity 184
thermistor 163, 166, 172
thermoplastics 189, 199, 202
three dimensional form see 3D
tone 49
tools 191, 194–6, 198, 203
   choosing 214–15
   cooking 232, 244
tempering 209
toughness 184, 209
toys 23, 27, 73, 141
transistors 164–5, 166
transmission mechanisms 101, 103, 109, 113, 122
trimmings 133, 195–7

U
user
   guide 97
   informing 77
   interface 68, 171, 173
   needs 31–2
   preferences 84
   support 96
   trip 72

W
wants 2–3, 75
warmth 2, 12–13
washing 125, 128
water 18
weather 14–15
weight 43, 184, 231
wind power 7, 96
winners 74
wood 6, 184, 186–7, 196
   cleaning 210
   fittings 205
   joints 206
worm and gear wheel 105

Acknowledgements

We are grateful to the following for permission to reproduce photographs and other copyright material:
(A = above, B = below, L = left, R = right, C = centre)

Ace Photo Agency 16B(Photo Library International); Andes Press Agency 34R, 72CB, 75L; Anglepoise–Nicky Smith Public Relations 120AL; Aviemore Photographic 99AL; BMW (GB) Ltd 7B, 22BL; Barnaby's Picture Library 6BL, 13AL & upper CR, 14B, 16C, 18C, 22BR, 26AR & lower C, 27B, 117BC; Biophoto Associates 42CL; John Birdsall Photography 33A, 42BL & BR, 72L, 186B, 187A, 213BR, 227AL, BC & BR, 229, 236AL, AR, upper CL & upper CR, 243A, CL & BR; Blundell Harling Ltd 117AR; Gareth Boden 4B, 5, 23B, 24B, 25; Bridgeman Art Library 33CL, CR & BR; British Motor Industry Heritage Trust/Rover Group 52; Neill Bruce Photographic 34A, 120C; J. Allan Cash Photolibrary 2A, 9BL & BC, 13 upper CL, 16A, 17C, BC & BR, 18A, 20CL, 21BR, 102A & BR, 107C, 115B, 222AL & AR; Chubb Alarms Ltd 24A & C; Chris Coggins 62–65 screen frames, 63L; Collections 71(Anthea Sieveking); Colorific! 145BR(Elisa Leonelli); Colorsport 99AC, 102CL & CR; Cotton Council International 123CL; Crafts Council 136AR(S. Bosence) & BL(Anne Sicher), 213A(J. Cleverly) & L(Martin Chetwin); Creda Ltd 96AL, AR & C; Paddy Cutts 8BR, 36, 61BL & BR, 68C, 69B, 72CA, 75CB, 100A, 101AR, 103B, 104A, 107A & B, 110AL, AR, CL & CR, 111, 113AL, AR, C & B, 114, 117AL, AC, CL & BL, 120AR, 135, 152A & inset, CL, CR & BR, 154, 184; Charlotte Deane 99BL, 108A, 120BL & BR; Dowdeswell Engineering Co Ltd 101B; Duracell UK–MDPR 152BL; Economatics (Education) Ltd 61AL & AR, 174BL, 176; Ford Motor Company 68; Werner Forman Archive 9AR, 11BL, BC & BR, 136AL & CL; Nick Given 115A; Halfords 103A; Robert Harding Picture Library 3, 6A, 12CB, 13 lower CR, 14A, 17AR, 21 upper CR, 22AR, CL & CR, 26AL, 33BL, 86, 117C(IPC Magazines), 136CR & BR, 145AL; Hasbro UK Ltd 27A, CL & CR; Michael Holford 99AR; Holt Studios International 123AC; Hotpoint 174AL; Houses & Interiors 29A; Hutchison Library 8AL, AR, L, C, CR & BL, 9AL & BR, 10, 11AR, 17CL; ICCE Photolibrary 43(Sue Boulton), 89(Sue Boulton), 99BR(Mark Boulton), 184C(Mark Boulton) & B(Mark Boulton); Richard Ingle 84B; Jaguar 98A; Kimberly-Clark Ltd 7C; Neville Kuypers Associates 218, 219, 222B, 223, 238; Lego UK Ltd 61CL & CR; Longman Logotron, screen image 65B; Longman Photographic Unit 4AL & AR, 17AL, 18B, 19, 21AL, AC & AR, 23AL & C, 35, 49, 57, 84A & C, 92, 98B, 100C & B, 109B, 123BL, BC & BR, 126, 127, 128, 130, 132A & B, 187B, 188, 189, 202, 203, 207, 209, 211, 212, 216, 217, 224, 226, 227AR, CR & BL, 230, 236 lower CL & CR, BL & BR, 242, 243CR & BL; Mansell Collection 42A; McDonald's Restaurants Ltd 34L; Milepost 92½ 108BL & BR; Modus Project, The Advisory Unit for Microtechnology in Education, screen image 64B; Oak Solutions, screen images 62A & B, 63A & B, 64A & L, 65A & L; PPL Ltd 102BL; Philips Components 69R; Picturepoint 20L & AC, 21L; Pifco Ltd 110B; John Plater 87; Potterton Myson Ltd 69AL & AC; Louise Rutherford 28B; Science Photo Library 162L(Malcolm Fielding, the BOC Group PLC), 174AR(Sheila Terry); Sleepeezee Ltd 120AC; Harry Smith Collection 109A; Sony United Kingdom Ltd 104B; Stanley Automatic Doors 162R; Still Pictures 123AL(Paul Harrison), 186AL(Mark Edwards), AR(John Maier), CL(Mark Edwards) & CR(Nigel Dickinson); Tony Stone Worldwide 6BR(Bryan Parsley), 12CR(David Woodfall), 13AR, 17BL(Annette Soumillard), 20R(Warren Jacobs) & CB, 21C & CR(Penny Tweedie), 123AR(Arnulf Husmo), 145AR; TRIP 12AR(Norman Price), CL(Helene Rogers), BL(Helene Rogers) & BR(V. Kolpakov), 13 lower CL(Eye Ubiquitous-Paul Scheult) & BR(Helene Rogers), 15(Bob Turner), 17CR, 22AL(Eye Ubiquitous-Mostyn), 26 upper C(Brian Gibbs) & B(Kanashev); Technical Blinds 177; Telegraph Colour Library 7A(Bavaria Bildagentur), 12AL(Guy Hurlebaus), 13BL(FPG–E. Jacobson), 29B(S. Benbow); Shirley Thompson 101AL; Tony UK Ltd 23AR; John Walmsley 32, 37A, 68A, 72R; Wind Energy Group 96B(Margaret Haynes) ZEFA 2B, 11AL & AC, 31, 37BL & BR, 67, 75AR & BR, 123C & CR, 162C, 174BR.

We are grateful to the following organizations for their assistance in setting up photographs:
Harlow Technical College 202(except A), 207, 209A, CL, CR & BL, 212; Supercast 202A.

We are grateful to the Mirror Syndication International for permission to reproduce the article 'Double alert on egg and chickens peril' by Julia Langdon and Gordon Hay in the Daily Mirror 12·01·89.